Columbia University

Contributions to Education

Teachers College Series

No. 359

AMS PRESS
NEW YORK

THE RELATIVE MERITS OF CONVENTIONAL AND IMAGINATIVE TYPES OF PROBLEMS IN ARITHMETIC

By

HARRY GROVE WHEAT, Ph.D.

Teachers College, Columbia University
Contributions to Education, No. 359

Bureau of Publications
Teachers College, Columbia University
NEW YORK CITY
1929

Library of Congress Cataloging in Publication Data

Wheat, Harry Grove, 1890-
The relative merits of conventional and imaginative types of problems in arithmetic.

Reprint of the 1929 ed., issued in series: Teachers College, Columbia University. Contributions to education, no. 359.
Originally presented as the author's thesis, Columbia.
Includes bibliographical references.
1. Arithmetic--Study and teaching (Elementary)
2. Arithmetic--Problems, exercises, etc. I. Title.
II. Series: Columbia University. Teachers College. Contributions to education, no. 359.

QA135.5.W45 1972 372.7'2 71-177643
ISBN 0-404-55359-1

Reprinted by Special Arrangement with Teachers College Press, New York, New York

From the edition of 1929, New York
First AMS edition published in 1972
Manufactured in the United States

AMS PRESS, INC.
NEW YORK, N. Y. 10003

ACKNOWLEDGMENTS

The writer of this report is indebted to the principals and teachers in the following schools, which coöperated in the investigations: the Emmons, Enslow, Guyandotte, Holderby, Johnson, Lincoln, Miller, Monroe, Oley, and Simms schools, Huntington, West Virginia; the Longfellow, Washington, and Whittier schools, Paducah, Kentucky; the Franklin, Grant, Jefferson, Lincoln, Longfellow, Sheridan, and Washington schools, Sheboygan, Wisconsin; the Knapp School, Racine, Wisconsin; and the training schools of the Grace Green Normal School, Dayton, Ohio, and of Marshall College, Huntington, West Virginia.

He is under special obligation to the following for assistance in giving the tests: Professors R. I. Roudebush and Roy C. Woods, of Marshall College; Miss Gertrude Roberts, Supervisor of Mathematics, Huntington, West Virginia, Public Schools; Miss Wilma Spear, Supervisor of Training, Grace Green Normal School, Dayton, Ohio; Superintendent L. J. Hanifan, Paducah, Kentucky; Principal F. S. McCall, Racine, Wisconsin; and Principals C. F. Lutze and E. H. Weisler, Sheboygan, Wisconsin.

He gratefully acknowledges his indebtedness to his dissertation committee, Professors J. R. McGaughy, Harold Rugg, and C. B. Upton, of Teachers College, Columbia University, for their assistance and criticism; to Dr. Harold Rugg, for the encouragement extended from the beginnings of the undertaking; and to Dr. John R. Clark, Dr. W. S. Elsbree, Dr. W. D. Reeve, and Dr. Helen Walker, of Teachers College, Columbia University, and his wife, Florence Bodey Wheat, for a critical reading of the manuscript.

H.G.W.

CONTENTS

CHAPTER PAGE

I. Introduction 1

II. The Purpose of Problem-Solving in Arithmetic . 8

III. The Method of the Investigations 17

IV. Differences in Pupil-Responses to the Shorter Conventional and Imaginative Problems . . . 39

V. Differences in Pupil-Responses to the Longer Conventional and Imaginative Problems . . . 63

VI. Similarities in Pupil-Responses 79

VII. The Influence of Time as a Factor 95

VIII. The Influence of Change 106

IX. Summary and Conclusions 116

CHAPTER I

INTRODUCTION

The investigations reported in this monograph deal with the responses of school children to arithmetic problems of the conventional type and to arithmetic problems of the imaginative type.

Definitions of Terms

The term "conventional" is used to describe the type of problem in arithmetic which is stated in the shortest, simplest, and most direct language possible. The conventional problem contains no mention of, nor allusion to, any element of the total situation under consideration—however interesting or appealing such element may be—other than those which may be described as the mathematical elements, or which require mathematical terms for their portrayal, or demand mathematical thinking for their complete understanding. The following are illustrations of problems of the conventional type:

1. Margaret spent $3.68 for handkerchiefs at 23 cents each, and gave one-fourth of them to her sister. How many did her sister get?
2. The fifth-grade class made 173 pop corn balls. They ate 35 of them and put the rest into sacks, putting 3 balls into each sack. How many sacks of pop corn did they have?

The term "imaginative" is used to describe the type of problem in arithmetic which contains elements of the total situation which are interesting and have an imaginative appeal, but whose presence is not germane to the problem under consideration nor necessary for an understanding of the mathematical elements of the situation presented. The imaginative problem is designed to "stimulate vivid imagination"[1] and to "engage the interest of the pupils."[2] The following are illustrations of problems of the imaginative type:

[1] Garry C. Myers, *Errors in Arithmetic,* p. 65. The Plymouth Press, Chicago, 1925.

[2] B. R. Buckingham and W. J. Osburn, *The Buckingham-Osburn Searchlight Arithmetic, Book One,* p. 1. Ginn and Company, Boston, 1927.

3. Margaret had been shopping all morning for Christmas presents. She had bought presents for her father and mother and brothers but could not decide what to get for her sister and several of her friends—there were so many things to pick from. Just then she saw some pretty handkerchiefs which were marked 23 cents each. These were just what she wanted, so she counted her money, found that she had $3.68, and spent all of it for handkerchiefs. She kept out one-fourth of the handkerchiefs to give to her sister, and gave the rest to her friends. How many did she keep out to give to her sister?

4. The boys and girls in the fifth grade wanted to make some money to buy presents for the Children's Hospital at Christmas time. They decided to make pop corn balls to sell. Altogether they made 173 pop corn balls, but while they were getting ready for the sale they ate 35 of them. They put the rest into sacks with 3 balls in each sack. How fat the sacks looked and how fast they sold. When all the sacks were filled they had how many sacks on the table for sale?

The Problem of the Study

The problem of the investigations which have been undertaken is a four-fold one:

1. To determine which type of arithmetic problem, the conventional or the imaginative, possesses the greater value as an exercise to aid in generalizing knowledge of the fundamental operations.
2. To determine the differences in degree of helpfulness of the two types of problems to pupils of slow, average, and superior attainments.
3. To determine whether the differences in degrees of helpfulness in individual cases are due (1) to differences in ability to read, (2) to differences in general intellectual ability, (3) to differences in carefulness of work, (4) to differences in reasoning ability in arithmetic, or (5) to differences in ability in the fundamentals.
4. To study the similarities in the responses of pupils to the two types of problems, and to determine the reasons for such similarities.

The Source of the Problem

The problem arises out of certain references in recent educational writings to an alleged superiority of arithmetic problems of the imaginative type, and out of a recently developed, though perhaps no more than apparent, tendency on the part of makers of textbooks to publish bulkier books in order to include many of the longer, and so-called more imaginative problems.

The importance of adding to the arithmetic problem of the usual, short, conventional type imaginative elements for the sake of making vivid the situation described was suggested by Wilson [3] in a report of the employment of specific methods of training pupils to read arithmetic problems. The methods employed were designed to train pupils to ask themselves and to answer for themselves questions about the situation described in a problem. For example,

> The pupils used the mere facts of the problem as the plot of a story. They read between the lines and supplied the details which made the problem real. They changed dull facts about buying and selling sugar into fanciful tales of lemonade stands.
>
> Such a problem as "What was the cost of 6 pounds of sugar at 8 cents a pound, 6 cans of milk at 5 cents a can, and 2 dozen eggs at 30 cents a dozen?" becomes for Louise a shopping situation in which her mother wants to bake a cake—unexpected guests—Louise must make a hurried expedition to the grocery. Her mother needs three things—eggs, milk, sugar. She says them over and over on the way so as not to forget the important items. The grocer quotes prices as she buys each article. . . . She gets home and reports to her mother the total expenditure, and her mother in turn, counts up to see if Louise has figured the problem correctly.

Exercises of the kind just described were reported to have led to a marked improvement in the ability of the pupils to read arithmetic problems and to deal with the situations which the problems presented.

It would seem at first glance that the series of exercises reported by Wilson makes a clear case for the employment of imaginative material in the statement of problems. It is to be noted, however, that the report at no point suggests that the work of including the imaginative material shall be the specific task either of the teacher or of the textbook, but that the pupils in every case shall be trained to provide for themselves their own imaginative material. Notwithstanding, Myers,[4] after quoting Miss

[3] Estaline Wilson, "Improving the Ability to Read Arithmetic Problems," *Elementary School Journal*, XXII, pp. 380-86.

[4] Garry C. Myers, *Errors in Arithmetic*, p. 64.

Wilson's findings, uses them in partial justification of his own conclusion that "a good problem, therefore, is one which leads the learner so to make it his experience as to render inevitable a right solution"—apparently assigning to the problem, but not to the pupil, the task of providing the imaginative elements.

Myers [5] believed that "most concrete problems in arithmetic are dry and tiresome," and less readily comprehended by pupils than similar problems "designed to stimulate vivid imagination." He designed a test containing six problems, each phrased in two ways. The two presentations of one of the six problems follow:

A girl caught 35 of her string of 135 beads when the string broke, and could find only half of the rest. How many beads did she have?

Bertha's daddy bought her a new string of 100 beads. While she was playing tag, the string broke. Poor Bertha! The beads scattered in every direction. She caught 48 of them before they fell, but she found only half of the rest. When she put the beads on another string she had only .. beads?

The six pairs of problems, arranged in mixed order, were solved by 513 persons. In five of the pairs of problems, the imaginative type was reported noticeably easier. Myers concluded that "the greater facility of this kind of problem [the imaginative] is attributable to its emotional appeal, to the clear-cutness of the images it elicits. There can be no question about the greater interest which such problems stimulate."

The findings and conclusions of Myers have been reported with favorable comment and recommendation by Morton [6] in his manual on the teaching of arithmetic, and by the Sub-Committee on Arithmetic of the 1924-25 Commission on the Curriculum of the Department of Superintendence of the National Education Association.[7] No question is raised in these reports as to whether the experimentation upon the relative merits of the two types of problems has been extensive enough or careful enough to warrant comment and recommendation conclusively favorable to either type.

Turning to a consideration of the newer textbooks in arithmetic as a source of the problem of the present study, one finds

[5] Garry C. Myers, *Errors in Arithmetic,* pp. 65-68.

[6] R. L. Morton, *Teaching Arithmetic in the Intermediate Grades,* pp. 306-7. Silver Burdett and Company, New York, 1927.

[7] *Third Yearbook,* pp. 94-95. The Department of Superintendence of the National Education Association, Washington, D. C., 1925.

in certain instances evidences of deliberate planning to enlarge the pupil's book by including within the course not merely more problems, but problems of longer and more imaginative type. Reference will be made at this point to two recent series of texts, namely, the *Standard Service Arithmetics*[8] and the *Searchlight Arithmetics*.[9]

The former series offers approximately 400 pages of printed material for the pupils of each grade. Counting in each case the pages of the Index, the books of Grades III, IV, and V offer for the consideration of the pupils 391, 443, and 414 pages, respectively. The authors lay special stress upon the fact that "one of the most striking features is the generous amount of space taken in presenting the work for each grade,"[10] and call particular attention to the "unusual amount of space devoted to pictorial and diagrammatic aids to understanding."[11] Considerable space is used for the presentation of such problems as the following:

3. The secretary kept the names of the children who were to serve on Committees. Four were to serve on the Lunch Room Committee, 2 on the School Ground Committee, 3 on the Program Committee, 1 on the Red Cross Committee, 2 on the Safety Committee, and 13 on the House Committee. How many names should the secretary have had on her lists? (Grade IV, p. 112.)

The latter series offers a total of 1,598 pages of material for the five grades, III to VIII, as follows: for Grade III, 358 pages, for Grade IV, 359 pages, for Grades V and VI, 491 pages, and for Grades VII and VIII, 390 pages. Problems like the following consume much space in the four books:

16. Lou had made a dainty paper chain. I think she made it in school. Anyway it was green, and red, and it was 15 feet long. Fred and Lou were draping it on the tree when in rushed the family pup and caught the end that was hanging down. He ran away with a piece of the chain. Then it was only 12 feet long. How many feet did the pup run away with? (Book I, p. 77.)

[8] F. B. Knight, J. W. Studebaker, and G. N. Ruch, *Standard Service Arithmetics, Grades Three, Four,* and *Five.* Scott, Foresman and Company, Chicago, 1927.

[9] B. R. Buckingham and W. J. Osburn, *The Buckingham-Osburn Searchlight Arithmetics, Books I-IV.* Ginn and Company, Boston, 1927.

[10] *Standard Service Arithmetics, Grade Three,* p. iv.

[11] *Ibid.,* p. vii.

The authors lay emphasis upon the fact that their problems "relate to almost every known aspect of childhood interest. They enter the fields of literature and fancy. Moreover, they have been written with the frank purpose of engaging the interest of the pupils. Why not?" ask the authors. "Must arithmetic be dull in order to be acceptable?"[12] The reader is confronted with the question whether a problem is necessarily dull if it does not enter the fields of literature and fancy.[13]

Myers,[14] in a recent review of the *Searchlight Arithmetics*, comments as follows:

> They have dared to break away from the traditional type of problem with the juice of human interest all squeezed out. They have taken account of the recent scientific evidence that long problems with interesting details are easier than are shorter ones which involve a like number of difficulties. . . . The lean textbook is sure to die, and fatter ones are just as certain to survive.

Questions Raised

As one reviews the meager literature upon the relative merits of the two types of problems under consideration in conjunction with the recent textbooks which expand their covers through the inclusion of problems of the longer type, he encounters throughout questions like the following:

1. Can the evidence upon the relative merits of the two types of problems be said to be scientifically determined?
2. Is the evidence which is presented based upon experimentation that is sufficiently extensive?
3. Is a problem which is stated in language void of imaginative details necessarily dull?
4. What kind of interest are problems in arithmetic designed to stimulate?
5. Whose is the task of providing the necessary imaginative material for the problem—the text's or the pupil's?

[12] *The Buckingham-Osburn Searchlight Arithmetics, Book I,* p. ix.

[13] The present writer does not intend to imply that the larger than usual size of the two series of texts named is due exclusively to the presence of longer problems. He wishes merely to impress the fact that the presence of the longer problems requires extra space in the texts named and is a reason for their larger size.

[14] *Elementary School Journal,* Vol. XXVIII, pp. 150-52.

In short, one encounters at every point—between the lines, to be sure—the question whether the elementary school is ready to abandon the terse, economical, compact, conventional form of expression in the language of its problems in arithmetic in favor of the form which is abounding in interesting details that are not germane to the situations to which the attention of the pupils must be called. The answer of the writings to which reference has been made in foregoing paragraphs is that the imaginative problem is the better because it is the easier. The task of the present study is (1) to decide what merit shall be assigned to ease of solution and (2) to determine which type of problem really possesses the greater merit and for what class of pupils such merit is helpful. The purpose of problem-solving arises for consideration in order to present a background for a report of the investigations which have been made upon the problem of this study.

The Limits of the Present Study

The present study will be limited to investigations of problem-solving activities which involve the use of the four fundamental operations of addition, subtraction, multiplication, and division in familiar situations.

Summary

The present study has been undertaken in order to determine which type of arithmetic problem—the conventional or the imaginative—possesses the greater merit, and, if any special merit is discoverable, to determine for what class of pupils such merit is helpful. The problem arises out of: (1) the claim of superiority for the arithmetic problem of the imaginative type which has been made in certain educational writings, and (2) the discernible tendency in pupils' textbooks to expand in order to include the longer problems. The study will not extend beyond investigations of problem-solving activities in the four fundamental operations.

CHAPTER II

THE PURPOSE OF PROBLEM-SOLVING IN ARITHMETIC

Consideration of the purpose of problem-solving in arithmetic requires attention to the nature and development of number ideas as well as to their proper employment in the problem-solving activity.

Number Facts Versus Number Ideas

One may look upon the learning of arithmetic as the learning of separate number facts, or one may view the process as the development in the mind of the learner of interrelated number ideas which may be applied by him both in clarifying the ideas he possesses and in developing ideas of higher orders. Viewed from the former angle, the learning of arithmetic is little more than systematic memorization. Viewed from the latter angle, the process is that of an active mind which not only learns the various number facts but also understands them and their relations and seeks possibilities of their applications.

Isolated Learning

There are many pupils in school who succeed in learning their arithmetic only as a mass of isolated and unrelated number facts. The number relations in the fundamental operations of addition, subtraction, multiplication, and division are learned only as so many facts to be remembered. Through constant drill and persistent effort, they succeed in acquiring skill in the operations, but fail to recognize withal the nature and the meaning of the operations. As a result, they learn to add, subtract, multiply, and divide with a fair degree of mechanical precision; they learn to perform such operations as they may be directed to perform; but they do not develop the ability to recognize the presence of these operations in the simplest of practical situations in which they may be found. In the course of time, however, they succeed in remembering that a statement which includes such terms as "how many," "altogether," "total," "sum," etc., requires addition; that a statement in which such terms as "take away," "left," "how many more," etc., are found requires subtraction; and so on. In other words, they learn to remember the various computations; they come to regard them as so many mechanical and meaningless performances; and, finally, they learn such of their applications as can be remembered by formula and rule.

Such learning is slow, tedious, and difficult. In such learning "five and three are eight" is a fact to be remembered quite apart from any relation it may have to "six and two are eight," etc., either in developing a general notion of the idea "eight" and of the relation of "eight" to the ideas "five," "three," "six," "two," etc., or in making clear the process of thinking things together in the development of the general idea of addition.

The Development of Ideas

On the other hand, there are pupils in school who learn their arithmetic as a system of related ideas which holds together and makes easy to remember the various facts of the combinations. To them "five and three are eight" is not merely a fact to be remembered, but a statement of a particular way in which a group of five and a group of three may be regrouped and a statement of the relation of the ideas "five" and "three" to the idea "eight." Learned in connection with the various ways a group of eight may be regrouped, "five and three are eight" becomes an idea of arrangement that connects in their thinking with "six and two are eight," "one and seven are eight," "two fours are eight," "five from eight is three," "eight is five more than three," etc. In other words, as the pupils give special consideration to the idea "eight" and then learn more about it than they had previously known, they learn several number facts in their proper relationships.

Likewise, the various facts of addition, for example, are learned as illustrations of, and while the pupils are giving special attention to, the mental process of thinking things together, that is, the general idea of addition. To be sure, the pupils need plenty of drill upon the various facts of addition, once they have been learned, in order that such facts will be ready at all times for immediate use. They need, moreover, and the school makes special provision for, plenty of practice in recognizing addition in a variety of practical situations. By such means the pupils not merely acquire reasonable skill in computation but also develop a better understanding of what the process of addition really is.

It is to be remembered that ideas of number and of number relations are not sense facts which the child can learn through observation; neither are they of the nature of factual experience which the teacher can pass along to her pupils by means of oral and written language. Like all other ideas, the ideas of number and of number relations develop in the mind of the individual who responds mentally in a suitable way to his experiences. Ideas grow in the mind; they are the products of mental activity. If the child is to be induced to develop number ideas, he must be given training in reacting in an orderly and systematic way to surrounding experiences. In this sense, learning number ideas is not to be thought of as the acquiring of new experiences. It is rather the process of organizing experiences—the activity of arranging experiences and of giving attention to their arrangement.

The Study of Arrangement

Illustration may be drawn from the methods of instruction resorted to in the early grades of the school. In the beginning, objects are used for the purpose of demonstration. Objects are presented, not for the purpose of calling attention to the objects themselves or to any of the concrete characteristics which they possess, but for the sake of directing attention to the ways the objects may be arranged. Since it is not possible at the outset to direct the attention of pupils to the various processes of grouping by means of oral direction or by means of the number symbols, objects in various arrangements are presented and these arrangements are pointed out for special study. For example, the fact that a group of eight is the same as a group of five and a group of three taken together is illustrated by eight objects arranged in the groups required. As the pupils study the arrangement, they are helped in their study by the language forms presented by the teacher—"five and three are eight" (orally), and $\begin{array}{r}5\\ \underline{3}\\ 8\end{array}$ (in writing)—which forms are used to describe the arrangement and are presented in order to serve as a means of fixing the arrangement and facilitating future thinking about it.

As the pupils develop in the use of the language of number, both the oral and the symbolic, there is less and less need to resort to objects in various groupings in order to call attention to arrangement. Finally, the objects may be dispensed with altogether. The various steps of progress in the ability of pupils to give attention to arrangement may be indicated thus:

1. Actual arrangement of objects by the teacher.
2. Actual arrangement of objects by the pupils.
3. Arrangement by pointing first to one group, then to the next.
4. Arrangement by looking first at one group, then at the next.

In (3) and (4), the pupils do not actually disturb the objects. They *think* the arrangements. Throughout, the language of number, that is, the oral statements and the written statements with the number symbols, is used to describe the actual and the imagined arrangements.

5. Attention to arrangement when no objects are present, employing the language of number to describe the arrangement which is now entirely a matter of thought.

Training in Abstraction

From the beginning the pupils are trained, with greater or less success, to be sure, in abstraction. As has been pointed out, objects are presented, not for study, but in order that their grouping may be observed. While the pupils must necessarily look at the objects, they are trained to recognize their arrangement. Color, shape, size, and other qualities of the objects must be neglected. Indeed, as the training progresses, the objects become hazier and hazier in the attention of the pupils, and the facts of arrangement stand out in clearer and clearer perspective. The time quickly comes when the pupils can give attention to ideas of arrangement when no objects are present to illustrate the arrangements under consideration.

Illustrative Practice Exercises

When the pupils have learned to *think* the arrangement of objects without actually disturbing them, it is but a further step to *think* the arrangement of objects when the objects are present only in imagination. For the latter purpose, problems are presented for their consideration.

Problems are often called "examples." The word "example" is the more fitting and exact term to use because the problems are in reality examples or illustrations of matters which are more important. Problems are often considered in arithmetic courses as the center around which everything else should revolve. Solving the problem and getting the answer occupy the center of the stage. The real purpose of the problem is to illustrate, clarify, and enlarge an idea to which the pupils have been giving attention.

In the problem, "John caught 5 fish and gave Henry 2. How many did John have left?", the pupils are presented with a situation which is not actually present to the senses. To be sure, the interest of the pupils is in the total situation; they have, however, in the number relations, that is, in the rearrangement of the objects, a very special interest. Heretofore, the pupils have had before them five objects, actually present, and have laid aside two, and observed three objects remaining. They have described the arrangement orally, thus, "two from five is three," and in writing thus, $\begin{array}{r} 5 \\ -2 \\ \hline 3 \end{array}$. In the problem relating to John's fish, the pupils are compelled to *think* the rearrangement of the five fish. They can give little or no attention to the fish as individual objects. If the numbers were larger, they could not visualize the objects except in a vague, general, and indefinite outline of the whole group. Without giving attention to the fish, the pupils give attention to their division into two groups, employing the language of arrangement, which they have developed in connection with the actual arrangement of objects, to describe the arrangement which is present only as a matter of thought. Because they have been guided through language to attend to the grouping of objects actually before them, they have learned to think in exact terms of this similar situation which is not actually before them.

Moreover, dealing with a situation not actually present enlarges the pupils' idea of arrangement. In the activities with objects, "two objects from five objects," for example, has been demonstrated repeatedly. The pupils have taken two crayons from five crayons, two books from five books, two pencils from five pencils, etc. Each succeeding activity has helped not only to make the fact "two from five is three" more definite but also to make more general the "take away" idea of subtraction. It is merely a further step in practice to take two fish from five fish and have three fish remaining, when the fish are present only in the pupils' imaginations.

The Language of the Problem

The language of the problem should be so clear as to leave no doubt in the child's mind as to what the situation really is. The language should not be difficult nor strange. Unusual terms have no place. Such statements are axiomatic and of common acceptance. It may be added, moreover, that it seems to follow logically that the teacher does not wish the pupils to give attention to the language of the problem, but to the arrangement which is described. Likewise, it seems to follow logically that the language of the problem ought not to attract attention to exciting or unusual phases of the total situation. It may add to the general enthusiasm for the problem of "John's fish" to describe at length John's fishing trip or the exciting adventures he experienced in catching the five fish; but it appears that it is not interest in John nor his adventures that is important. The language of the problem should point unmistakably to the simple fact that John had five fish and gave away two fish. It is the number situation—the fact of arrangement—which should command the attention. Whether embellishment of the language of the problem aids or not in directing attention to the fact of arrangement is the problem which the present study will attack.

Solving the Problem

The "problem" of "John's fish" is not really a problem. Within it there is no element of "perplexity, confusion, or doubt," nor does it present to the mind any "unexpected, queer, strange, funny, or disconcerting" elements which Dewey [1] has described as furnishing the starting point of reflective thought. It can not be classed

[1] John Dewey, *How We Think*, p. 64. D. C. Heath and Company, 1910.

under the heading of the "genuine problem" which Dewey speaks of as existing in "whatever—no matter how slight or commonplace in character—perplexes and challenges the mind so that it makes belief at all uncertain." [2] True, the pupil may be perplexed, in confusion, or in doubt when such a statement is presented to him, as pupils often are when "problems" in arithmetic are presented to them. In such a case, the problem is not arithmetical, but one of what the language of the statement is really trying to present. If the language of the "problem" of "John's fish" is clear, there is no problem. The pupil either recognizes in the statement the general idea of subtraction, or he does not. In either case, he engages in no reflective thinking in so far as the arithmetical situation of arrangement is concerned.

Definitions of Problem and Problem-Solving

Throughout our discussions, the terms "problem" and "problem-solving" will be retained, since they are in common use. By "problem" is meant "practice exercise," and by "problem-solving" is meant the mental process of recognizing the general ideas of addition, subtraction, multiplication, and division.

Ease of Recognition as a Criterion of Merit

A practice exercise should be difficult enough to command the pupil's attention, otherwise his practice with it has reached the point where the returns are greatly diminished. On the other hand, a practice exercise should be easy enough for uninterrupted application to the task at hand, that is to say, the "problem" should not be so difficult as to retard a ready recognition of the idea of arrangement which its statement is intended to describe. The type of "problem," therefore, which provides up to the point of diminishing returns the readier recognition of the fundamental operations involved will be regarded as the one possessing the greater merit for practice.

Purposes of Problem-Solving Activities Distinguished

The foregoing discussion of "problems" as practice exercises suggests a distinction between the type of arithmetic problem, so-

[2] John Dewey, *How We Think*, p. 9

called, which deals with familiar situations *during* the process of learning the meaning and use of the fundamental operations and the type of arithmetic problem which involves the use in new situations of the fundamental operations *after* they have been learned. Problems, that is, practice exercises, of the former type have been discussed in the present chapter. Problems of the latter type need to be set off in sharp distinction for the reason that the present investigation will not be concerned with them.

In the later grades of the elementary school, after the pupil has sufficiently mastered the fundamental operations and has become sufficiently aware of their meaning and use, he is introduced to such situations as savings and investments, insurance, business practices, mensuration, and the like. Here he studies the new situations, each of which possesses among others a mathematical side, and considers various problems which arise out of these situations. In problems of the former type, referred to in the foregoing paragraph, the pupil does not study the situation, because it is a familiar one, or one easily apprehended. In problems of the latter type, the pupil's chief concern is the study of the situation.

Limited Extent of the Present Study

With problems of the latter type the present study of the relative merits of conventional and imaginative problems will not deal. This study will concern itself only with the responses of pupils to problems—or practice exercises—involving the use of the four fundamental operations in familiar situations. The findings which will be reported may not, therefore, be considered as being applicable to the construction of problems in the situations which comprise in large measure the content of the course in arithmetic in the later years of the school. We shall be concerned here only with the type or types of problem practice exercises which are intended to aid the pupil in generalizing and extending his knowledge of the meaning and use of the fundamental operations.

Summary

The present chapter defines the learning of arithmetic in its early stages, not as the mere memorization of number facts, but as the development of ideas of number and number relations. Such ideas result from the giving of attention in a systematic way to arrangement: first, to objects in actual arrangement; second, to objects arranged in thought; and, finally, to facts of arrangement when the objects are present only in imagination. "Problems" are thus defined as practice exercises, and "problem-solving" as the activity of recognition of ideas of arrangement made discoverable by the language which defines a familiar situation. The type of practice exercise which provides the readier recognition of arrangement to the point of diminishing returns is held to be of the greater merit as a means of training. Such definitions of "problem" and "problem-solving" set off in sharp contrast the problem-solving activities of arithmetic in its later stages with which the present study will not be concerned.

CHAPTER III

THE METHOD OF THE INVESTIGATIONS

The Tests

The method of the investigations was the administration of five series of tests. These have been designated and will be referred to as Problem Test I, Problem Test II, The Collateral Tests, Problem Test IV, and Problem Test V, Parts A and B. They will be described in order.

Problem Test I

In order that a clear impression may be conveyed as to the nature of the test, Problem Test I and the directions for its administration are reproduced in full. An examination of the test shows that it is composed of ten pairs of problems, one conventional and one imaginative problem in each pair. The pairing is indicated in numbered blocks at the beginning of the test. For example, problem 9 pairs with problem 14, problem 15 with problem 10, and so on. The odd-numbered problems are of the conventional type, and the even-numbered problems are of the imaginative type.

Each problem is a two-step problem. The ten pairs of problems present ten combinations of the four fundamental operations, as follows:

Problems 9 and 14 are addition-addition problems.
Problems 15 and 10 are addition-subtraction problems.
Problems 7 and 4 are addition-multiplication problems.
Problems 1 and 6 are addition-division problems.
Problems 11 and 16 are subtraction-subtraction problems.
Problems 3 and 12 are subtraction-multiplication problems.
Problems 19 and 2 are subtraction-division problems.
Problems 13 and 18 are multiplication-multiplication problems.
Problems 17 and 19 are multiplication-division problems.
Problems 5 and 20 are division-division problems.

PROBLEM TEST I

9	15	7	1	11	3	19	13	17	5		
14	10	4	6	16	12	2	18	8	20		

	A.	S.	M.	D.	T.
Conv.					
Imag.					

Name.....................................Age last birthday.....years

Birthday.............................School................Grade.....

Month Day

1. How many hours must a boy work in order to pay for a tennis racket at $5.50 and a ball at 75 cents, if he makes 25 cents an hour? — *Work problem 1 here*

2. The boys and girls in the fifth grade made 173 pop corn balls at Christmas time. They ate 35 of them and put the rest in sacks with 3 balls in each sack. How fat the sacks looked and how fast they sold! When all the sacks were filled they had how many sacks on the table for sale? — *Work problem 2 here*

3. A grocer bought 30 dozen of eggs at 37 cents a dozen, and sold them at 43 cents a dozen. How much did he make? — *Work problem 3 here*

4. On our street there are 12 girls. Louise and Mary wished to form a club and the rest of the girls wished to join when they heard the plans. To belong to the club each girl was asked to pay 20 cents for her fees, 10 cents for the pleasure box, and 25 cents for charity. Each girl has to pay these dues every month. What does it cost the 12 girls each month to belong to the club? — *Work problem 4 here*

5. John spent 60 cents for marbles at 2 cents each, and divided them among 6 boys. How many did each boy get? — *Work problem 5 here*

6. Sam delivers groceries for Mr. White after school hours, and Mr. White pays him 20 cents an hour. Sam has decided to buy a football which will cost $1.25 and a pair of football shoes which will cost $2.75. How many hours will Sam have to work delivering groceries for Mr. White in order to buy the football and football shoes?

Work problem 6 here

7. Billy sold 55 papers and James sold 61. How much did they both make if they sold the papers at 4 cents each?

Work problem 7 here

8. The girls in the fifth grade were trying to make some money to buy new victrola records which cost 75 cents each. They had talked of many plans to make money until at last they agreed to make candy and sell it. They made the candy and sold 20 pounds of it at 60 cents a pound. How many records were they able to buy with the money they made?

Work problem 8 here

9. Jane gave $20 toward buying a radio, Harry gave $35, and John gave as much as Jane and Harry. How much did they all give?

Work problem 9 here

10. Nancy saw a pretty red raincoat in a window down town which cost $4.50. She saw some galoshes to match which cost $2.00. Nancy has been making some money by taking care of Mrs. Smith's baby on certain afternoons. She has saved $5.25. How much more will she have to save before she can buy the raincoat and galoshes?

Work problem 10 here

11. Joe made $45.80, and spent $10.80. How much does he need to make a $75 payment on a Ford?

Work problem 11 here

12. The Boy Scouts of Troop No. 13 wanted to make some money to build a camp. They knew that every loyal supporter of the high school football team would want to show his colors at the next big game, so they bought 600 pennants at 18 cents each. Just before the big game they sold all the pennants for 25 cents each.

Work problem 12 here

How much money did they make for their camp?

13. How much did a boy earn who worked 8 hours a day for 26 days at 40 cents an hour?

Work problem 13 here

14. Mary, June and John had been saving their money for weeks to buy their mother's Christmas present. Mary had saved $1.20 and June had saved 50 cents. John sold papers and had been able to save more than the girls. In fact he had saved as much as Mary and June together. How much money did they all save for mother's Christmas present?

Work problem 14 here

15. Mary has saved $6.20. How much more does she need to pay for a hat which costs $3.50 and a dress which costs $5.25?

Work problem 15 here

16. Mary had been carefully saving her nickels and dimes ever since she started to school in the first grade. She was now a big girl in the fourth grade and was very proud of the $45.80 in her book showing how much she had of her own in the bank. One day her father fell and was badly hurt. For weeks he was not able to go back to his work, and her mother had to draw out and spend $9.80 of her money. How much will Mary have to save in order to have $50 in the bank?

Work problem 16 here

17. Frank sold 45 dozen ears of corn at 30 cents a dozen, and with the money he bought young chickens at 50 cents each. How many chicks did he buy?

Work problem 17 here

18. Dora's big brother Tom wished to earn his own spending money. He lives near a large mountain hotel so he goes there to work. All through the bright fall weather he cleaned windows in the evenings after school. He gets 15 cents for cleaning a window. If he was able to wash 4 windows in an hour and if he worked 3 hours each evening how much did he earn each day after school?

Work problem 18 here

19. A farmer had 931 bushels of oats. He sold 485 bushels and put the rest into sacks, 2 bushels in each sack. Now how many sacks of oats did he have?

Work problem 19 here

20. One day Rose went to the Day Nursery to play. She thought it would be nice to take something to the children there. On her way she saw some red balloons and made up her mind to buy them. They were 4 cents each, and she had 80 cents which she spent for them. If there were 10 children in the nursery how many could she give each child?

Work problem 20 here

Have you tried all the problems? Turn back and make sure that you have not missed trying them all.

DIRECTIONS FOR GIVING PROBLEM TEST I

Special precautions should be taken throughout the test to prevent communication and to secure independent work.

Be sure that each pupil has a well-sharpened pencil.

Distribute the folders face up, and tell the children not to handle them until you tell them to. Distribute no other paper.

Have the children fill in the data called for at the middle of the first page. Assist them if necessary. Then read the following directions, using a folder to illustrate what you direct the pupils to do:

"In this folder (indicate front, middle, and back pages) you will find some problems in arithmetic. Opposite each problem is a blank space in which you are to work it. Do not work on any other paper, but put all the work for each problem in the blank space where it tells you.

"When I tell you to begin, work the first problem, then work the second, then the third, and so on. Try to work all the problems. Do not talk. Do not hurry. You will have plenty of time. Try to work all the problems. Do you understand?

(Be sure that all understand.)

"Begin."

As soon as a pupil finishes, collect his paper, but *be sure that he has tried all the problems.* During the test, pass among the pupils from time to time, and suggest to any who may have skipped any problems that they be sure to try the ones skipped.

Any written work, whether right or wrong, constitutes a trial. Give no encouragement other than to say, "Try this one," *if nothing has been written in any given blank space.*

The Single Variable

In the construction and the giving of this test, the effort was made to distinguish the various controlling factors and to keep them all constant save one and to measure that one. The factors which may be recognized in an examination of the test are: (1) the problem situations, (2) the operations involved, (3) the difficulty of the parts, (4) the time, and (5) the manner of stating the problems. The first four factors are theoretically constant; the fifth varies.

The situations in the twenty problems of the test are those which are familiar to boys and girls in the fifth grade of the elementary school. They are the kind of situations with which such pupils have become familiar in part through actual experience and in part through contact in the daily problem practice in their work in arithmetic. They are for the most part the situations of earning money, buying and selling, saving money for a purpose, and of child activity. They were selected by the writer and by members of his class in the psychology of arithmetic from lists of problems actually in use in the fifth and sixth grades under the instruction of the members of the class.

The operations involved have already been named. They are familiar to fifth-grade children for they are the operations for which specific instruction and practice have been given. No fractions, percentages, or decimals, other than those involved in dollars and cents, have been included. Pupils who have passed into the work of the fifth grade can add, subtract, multiply, and divide and they are able to recognize these operations in simple situations. In order to introduce the element of difficulty, these operations were combined in the test into two-step problems. It is more difficult to recognize addition, for example, when it appears in a situation with another, or a like operation, than it is to recognize addition when it appears alone. Moreover, in order to provide in the test recurrence of opportunity to recognize each operation, it was provided in each set of ten problems that addition, subtraction, multiplication, and division each appear five times.

The ten problems in each set, that is, the conventional and the imaginative, were considered of equal difficulty. This is not actually the case, as the results which will be presented in the following chapter will show. The results, however, will show the problems of each set to be sufficiently the same in difficulty to justify the giving of the same score to each problem. Moreover, each problem in either set is paralleled by a problem in the other which is for all practical purposes its equivalent with the single exception of manner of statement.

The directions indicate a possible variation in the time which each pupil tested would devote to each set of problems. In the present test the pupils were given all the time necessary to try all the problems. In actual schoolroom practice, opportunity is provided for deliberation in the solving of problems. In the problem-solving activities of the school, a trial is considered a trial, whether the pupil spends one minute or ten. Time to try all the problems in each set was provided. Thus, theoretically, the factor of time may be considered as constant. In Problem Test V, the time element will be given further consideration.

The contrasting manner of statement has been discussed in Chapter I. The odd-numbered problems are stated tersely; the even-numbered problems present in their statement certain imaginative elements not strictly germane to the problems themselves. It may be objected that the imaginative problems are not greatly the contrast of the conventional due to the fact that none succeeds in presenting a situation of great emotional appeal. There is in none of the imaginative problems much that is extra in statement to stir the imagination. To compare the responses of pupils to the two sets of problems, it may be argued, one ought to expand the ten imaginative ones to twice or three times their present length so that more of the "story" element would be present in each. In answer, it is to be pointed out that the ten imaginative problems consume 68 printed lines and the ten conventional ones but 24 printed lines, and that to expand the imaginative problems further would take them beyond the limits of practical usage. The imaginative problems are quite as long as the usable textbook in arithmetic will permit.

Method of Scoring

Since the purpose of the test was to discover the relative ease of recognition of the fundamental operations in the two sets of problems, account was taken only of the times the operations were recognized in the various problems. Following are six typical responses to problem 1:

```
  (1)                  (2)                    (3)
  5.50           .25) 5.50                25) 75
    75                ----                    --
  ----                  22                     3
  6.25

            (4)                          (5)
  .25) 5.50    25) 75                    5.50
       ----        --                     .75
         22         3                    ----
                                     25) 6.25
                                         ----
                                           25
                      (6)
              .25) 5.50    25) 75
                   ----        --
                     22         3
                      3
                     --
                     25
```

In solution (1), addition is recognized. One point is scored. In solutions (2) and (3), division is recognized. In either case one point is scored. In solution (4), division is recognized twice, but since the recognitions are similar, one point is scored. In solutions (5) and (6), both addition and division are recognized. In either case two points are scored and credit is given for the problem. No account was taken of inaccuracy in computation.

Problem Test II

Problem Test II was a repetition of the test given by Myers which is reported in his *Errors in Arithmetic*, pp. 65-67. The least difficult pair of Myers' problems was omitted. The test and the directions for giving it are reproduced on the accompanying pages.

PROBLEM TEST II

1. Paul Kraft had saved his money all winter so he could go to some of the big league games. He had saved $3.00. From this he bought a baseball for 60 cents. The rest he was going to spend on the games. "Let me see," he said to himself, "it will cost me 5 cents car fare each way, and 50 cents

to get in to see the game. With the money I have left I can see .... games."

2. Bertha's daddy bought her a new string of 100 beads. While she was playing tag, the string broke. Poor Bertha! the beads scattered in every direction. She caught 48 of them before they fell, but she found only half of the rest. When she put the beads on another string she had only .... beads.

3. Laura, whose grandmother gave her $1.50, saved 5 cents each school day from her lunch money. How many school days before Laura will have enough money to buy a book which costs $2.50?

4. If Rose spends two weeks' wages every month, earns $2.75 each Saturday, 50 cents for working from 5 to 7 o'clock in the evening of each school day, and 75 cents for each Sunday morning, how many months will it take her to save $60?

5. Little Betty wanted to surprise her mother on Christmas. She knew that her mother needed a good pair of scissors. The old ones would not cut well. One day Betty went with her daddy to a big store to see what a good pair of scissors would cost. She saw one for $2.00. Betty had 95 cents which her grandmother had given her. "I shall save 5 cents every day from my school money." After .... school days she bought the scissors, which lay at the foot of the Christmas tree on Christmas morning marked, "To my dear mother, from Betty."

6. After traveling 160 miles, a man has 4 gallons of gas left in his automobile. How many miles did he get to a gallon of gas if he bought 8 gallons on the way and had 6 gallons when he started?

7. A girl caught 35 of her string of 135 beads when the string broke, and could only find half of the rest. How many beads did she save?

8. A boy who had $2.55 bought a baseball for 75 cents. How many ball games could he see for the rest if he had to pay 50 cents for tickets to the game and 5 cents car fare each way?

9. The baby had been sick, mother was not well, the doctor's bill was $48, and the rent had to be paid. Helen was just old enough to earn some money, but she did not want to quit school. She found a job at helping a neighbor lady in the house on Saturdays all day and from 5 to 7 in the evening on school days. For Saturday she got $2.50, and for each school evening 50 cents. Every Sunday afternoon she kept Mrs. White's baby. For this she earned one dollar a Sunday. Helen had to have all she could earn two weeks out of every month to buy her clothes and school supplies. The rest she gave to pay on the doctor's bill. In .... months Helen paid off that bill.

10. Last summer Agnes Purdy, her brother Archie, and their parents took a trip in their Ford. Archie measured the gasoline when they started. "We have eight gallons," he told his father. At the end of the day he found

4 gallons of gasoline in the tank. They had bought 6 gallons at a station on the way, and had traveled 150 miles. Agnes told her mother that they had made .... miles to a gallon that day.

DIRECTIONS FOR GIVING PROBLEM TEST II

Special precaution should be taken throughout the test to prevent communication and to secure independent work.

Be sure that each pupil has a well-sharpened pencil.

Announce that you will presently pass out some problems in arithmetic. Have each pupil get ready by preparing a sheet of his tablet paper, as follows:

Name................................ School................ Grade......

Tell the pupils that there are ten problems for them to solve; that they are to put all of their work on the blank sheets upon which they have written their names, etc.; and to be sure to number each problem as they solve it.

Permit them to use back of blank sheet if necessary.

Distribute the problem sheets. Say to them:

"On this page (indicate) you will find ten problems in arithmetic. When I tell you to begin, work the first problem, then the second one, and so on, until you have worked them all. Do not talk. Do not ask questions. Do not hurry. You will have plenty of time. Do not mark on this sheet (indicate problem sheet). Try all the problems. Do you understand? (Be sure that all understand.)

"Begin."

As soon as a pupil finishes, collect his paper and the problem sheet. Keep separate. Be sure that the pupil has tried all the problems before collecting his paper. During the test, pass among the pupils from time to time, and suggest to any who may have skipped any problems that they be sure to try the ones skipped.

Any written work, whether right or wrong, constitutes a trial. If the pupil has written nothing for any given problem, give no encouragement other than to say, "Try this one."

Method of Scoring. In scoring this test only correct solutions were considered. The object was not to compare individual scores of pupils, but rather the total responses of many pupils.

The Collateral Tests

The collateral tests accompanied Problem Test IV and were given to the same pupils. They were the Burgess Scale for Measuring Ability in Silent Reading, Picture Supplement I; the Otis Self-Administering Test of Mental Ability, Intermediate Examination; selected parts of the Cleveland Survey Arithmetic Tests, Form I; and the Stone Reasoning Test, Form I. The object of the collateral tests was to provide a means of seeking the source of such difference in case any marked difference became evident in the relative merits of the conventional and imaginative problems.

The Burgess Scale for Measuring Ability in Silent Reading was selected as one of the collateral tests because it gives a ready and easily interpreted measure of the pupil's ability to grasp the essential thought in a short paragraph. The paragraphs of the scale compare favorably in length with the usual written problem in arithmetic. Moreover, to secure a mark of "right" on a paragraph of the scale the pupil must read the entire paragraph with reasonable care and must refrain from guessing. This the pupil must be schooled in doing if he is able to respond successfully to the printed problem exercises in his work in arithmetic. Thus the scale was considered as appropriate for providing a measure of one of the qualities necessary for success in solving problems in arithmetic.

The Otis Self-Administering Test of Mental Ability was selected because of its ease of administering and because pupils seem to experience little difficulty in understanding the directions to be followed. The test provides a measure of a pupil's standing relative to that of his classmates in keenness of insight, quickness in grasping a point, carefulness in work, and perseverance—all qualities necessary for success in carrying on the practice exercises with written problems in arithmetic.

While ability in computaticn may not be a help in interpreting the situation presented in an arithmetic problem, the lack of it is frequently a hindrance. A measure of this ability was needed. Accordingly, Sets A, B, C, D, E, F, G, and I of the Cleveland Survey Arithmetic Exercises were chosen as a test. They measure the pupil's skill in the simpler and slightly more complex processes in the four fundamental operations. The processes involved in the eight sets indicated are on the whole the processes involved in the necessary computations of Problem Test IV.

The Stone Reasoning Test was selected because it provided a fairly reliable measure of what is generally designated as reasoning ability in arithmetic. Hunkins and Breed[1] have reported the

[1] R. V. Hunkins and F. S. Breed, "The Validity of Arithmetical Reasoning Tests." *Elementary School Journal*, XXIII, pp. 453-66.

scores obtained by this test to be the least erratic and in closest agreement with the composite scores obtained by its use in combination with six other reasoning tests.

The collateral tests were administered and scored according to the sets of directions which accompany them. The score given to the pupils' responses on the Stone Reasoning Test was based upon the number of right answers. The computations in this test are so simple as to make the differences between scores based upon the right answers obtained and those based upon right principle only slight in amount and few in number. A special "index of carefulness" score was computed for the responses to the Otis Test of Mental Ability, and a composite score was developed from the response to the eight sets of the Cleveland Tests in the Fundamentals.

Index of Carefulness

In giving the tests designated as Problem Test I and Problem Test IV, it was noticed that many pupils were careless in their attacks upon the problems, and would skip about from problem to problem, leaving one before it was completed to attack another. Thus through failure to concentrate, or through a dissipation of effort, many pupils failed to do as well as they might, had they attacked the problems consistently and in succession.

The responses to the Otis Test of Mental Ability provided a means of securing a measure of each pupil's carefulness of attack upon the work at hand. The test contains seventy-five questions which are to be attacked in order. The pupil who secures a high score, other things being equal, is the one who gives careful attention to the questions one at a time and in order. The pupil who skips about from question to question, seeking the ones which appear easy, will succeed in securing only a low score. Accordingly, the per cent the number of questions answered correctly is of the number tried, including the number skipped, is a fair index of a pupil's carefulness of attack. Such an index of carefulness was computed for each pupil tested. The number of the highest numbered question attacked was taken as the number attempted; rather, as the number given some attention. This was divided into the total number of questions scored "right."

The Composite Score in the Fundamentals

In order to secure a composite score for the eight sets of the Cleveland Survey Test which were used, it was necessary first to equate the various sets. The method used is the one described by Counts,[2] which takes the average time required to work an example in one of the sets as a unit, assigns a value of 1.0 to each example worked in that set and, accordingly, equates the examples worked in the remaining sets.

In Table 1 the system of weights is shown and the method of deriving them is indicated. The scores used are those made by the 300 6B children tested. In order to indicate the equating of the weights both on the basis of difficulty and on the basis of the time allowances, the full procedure is shown.

TABLE 1

Derivation of System of Weights of Cleveland Test Scores

Set	A	B	C	D	E	F	G	I
Medium 6B Score	16.5	13.3	12.4	12.0	4.2	5.4	4.1	0.8
Time allowance in seconds	30	30	30	30	30	60	60	60
Score for 30 seconds	16.5	13.3	12.4	12.0	4.2	2.7	2.05	0.4
Weight relative difficulty	1.0	1.24	1.33	1.375	3.93	6.12	8.04	41
Weight after equating time allowance	1.0	1.24	1.33	1.375	3.93	3.06	4.02	20.5
Approximate weight used in computations	1	1¼	1⅓	1⅜	4	3	4	20

In the first horizontal column of the table are the average 6 B scores for eight sets; in the second is the time allowance in seconds; in the third is the average score per thirty seconds for each set; in the fourth are the measures of relative difficulty; and in the fifth and sixth are the weights after equating the time allowances.

[2] George S. Counts, *Arithmetic Tests and Studies in the Psychology of Arithmetic*, pp. 26-28. University of Chicago Press, Chicago, 1917.

Using the weights indicated, each pupil's scores in the eight sets were equated, and the summation of the equated scores was set down as his individual composite score.

Problem Test IV

In every respect—the form of test, method of administration, method of scoring the results, etc.—with the single exception of the manner of stating the conventional problems along with a slight expansion of the statement of problem 2, Problem Test IV is the same as Problem Test I.

PROBLEM TEST IV

9	15	7	1	11	3	19	13	17	5		
14	10	4	6	16	12	2	18	8	20		

	A.	S.	M.	D.	T.
Conv.					
Imag.					

Name.................................Age last birthday.......years
Birthday ....School..............Grade........
Month Day

1. Henry makes 25 cents every hour he works. He is saving his money to buy a tennis racket and a ball. He wants to know how many hours he will have to work to be able to pay for them, so he finds out how much they will cost. The tennis racket costs $5.50, and the ball costs 75 cents. How many hours will Henry have to work in order to pay for them?

Work problem 1 here

2. The boys and girls in the fifth grade made 173 pop corn balls at Christmas time. They wanted to sell them and make some money to buy presents for the Children's Hospital. They ate 35 of the pop corn balls and put the rest into sacks with 3 balls in each sack. How fat the sacks looked and how fast they sold! When all the sacks were filled they had how many sacks on the table for sale?

Work problem 2 here

3. A grocer bought a crate of eggs. He paid 37 cents a dozen for them, and he decided to sell them at 43 cents a dozen, since he wanted to make something on each dozen he sold. How much did he make on 30 dozen of eggs?

Work problem 3 here

4. On our street there are 12 girls. Louise and Mary wished to form a club and the rest of the girls wished to join when they heard the plans. To belong to the club each girl was asked to pay 20 cents for her fees, 10 cents for the pleasure box, and 25 cents for charity. Each girl has to pay these dues every month. What does it cost the 12 girls each month to belong to the club?

Work problem 4 here

5. John had 60 cents which he spent for some marbles. For each marble he had to pay 2 cents. When he had spent all his money, he counted the marbles. He had 6 sacks in his pocket, so he put the same number of marbles into each sack. How many marbles did he put into each sack?

Work problem 5 here

6. Sam delivers groceries for Mr. White after school hours, and Mr. White pays him 20 cents an hour. Sam has decided to buy a football which will cost $1.25 and a pair of football shoes which will cost $2.75. How many hours will Sam have to work delivering groceries for Mr. White in order to buy the football and football shoes?

Work problem 6 here

7. Billy sold 55 papers and James sold 61. They sold the papers at 4 cents each. When they were through they wanted to find out how much they both had made, so they put their money together and counted it. How much money did they have altogether?

Work problem 7 here

8. The girls in the fifth grade were trying to make some money to buy new victrola records which cost 75 cents each.

Work problem 8 here

They had talked of many plans to make money until at last they agreed to make candy and sell it. They made the candy and sold 20 pounds of it at 60 cents a pound. How many records were they able to buy with the money they made?

9. Jane, Harry, and John were talking about buying a radio. Jane said she would give $20.00 to help pay for it, and Harry said he would give $35.00. John counted up how much Jane and Harry would give. "I will give as much as you both," he said. How much did they all say they would give?

Work problem 9 here

10. Nancy saw a pretty red raincoat in a window down town which cost $4.50. She saw some galoshes to match which cost $2.00. Nancy has been making some money by taking care of Mrs. Smith's baby on certain afternoons. She has saved $5.25. How much more will she have to save before she can buy the raincoat and galoshes?

Work problem 10 here

11. Joe made $45.80, and he spent $10.80 of it. When he counted up what he had left, he said, "I have to pay $75.00 on my Ford and I do not have enough to pay it." How much more money did he need?

Work problem 11 here

12. The Boy Scouts of Troop No. 13 wanted to make some money to build a camp. They knew that every loyal supporter of the high school football team would want to show his colors at the next big game, so they bought 600 pennants at 18 cents each. Just before the big game they sold all the pennants for 25 cents each. How much money did they make for their camp?

Work problem 12 here

13. William works 8 hours every day and he makes 40 cents an hour. "That is a good day's wages," he said, as he figured up how much he made in a day. Last month William worked 26 days. How much did he make last month?

Work problem 13 here

14. Mary, June and John had been saving their money for weeks to buy their mother's Christmas present. Mary had saved $1.20 and June had saved 50 cents. John sold papers and had been able to save more than the girls. In fact he had saved as much as Mary and June together. How much money did they all save for mother's Christmas present?

Work problem 14 here

15. Mary wanted to buy a hat which cost $3.50, and a dress which cost $5.25. Mary counted up how much they both would cost and then looked in her pocketbook to see how much money she had. She had only $6.20. "I do not have enough money to buy them," she said. How much money did she need?

Work problem 15 here

16. Mary had been carefully saving her nickels and dimes ever since she started to school in the first grade. She was now a big girl in the fourth grade and was very proud of the $45.80 in her book showing how much she had of her own in the bank. One day her father fell and was badly hurt. For weeks he was not able to go back to his work, and her mother had to draw out and spend $9.80 of her money. How much will Mary have to save in order to have $50.00 in the bank?

Work problem 16 here

17. Frank sold 45 dozen ears of corn and got 30 cents a dozen. When he counted up how much money he had made, he said, "I will buy some young chickens with this." The young chickens cost 50 cents each. How many could he buy?

Work problem 17 here

18. Dora's big brother Tom wished to earn his own spending money. He lives near a large mountain hotel so he goes there to work. All through the bright fall weather he cleaned windows in the evenings after school. He gets 15 cents for cleaning a window. If he was able to wash 4 windows in an hour and if he worked 3

Work problem 18 here

hours each evening, how much did he earn each day after school?

19. A farmer had 931 bushels of oats. He sold 485 bushels. He still had a good many bushels left, so he decided to put what he had left into sacks. Each sack held 2 bushels. How many sacks of oats did he have?

Work problem 19 here

20. One day Rose went to the Day Nursery to play. She thought it would be nice to take something to the children there. On her way she saw some red balloons and made up her mind to buy them. They were 4 cents each, and she had 80 cents which she spent for them. If there were 10 children in the nursery how many could she give each child?

Work problem 20 here

Have you tried all the problems? Turn back and make sure that you have not missed trying them all.

In this test, the "conventional" problems are not strictly of the conventional type. Their manner of statement is of about the same length as that of the imaginative problems. They differ from the imaginative problems, however, in the respect that the extra wording used in their statement is used to describe more fully and to indicate more specifically what we may term the strictly mathematical elements of the problem situations. For example, the sentences, "so he finds out how much they will cost" in problem 1; "since he wanted to make something on each dozen he sold" in problem 3; "when he had spent all his money, he counted his marbles" in problem 5; and similar clauses in the other odd-numbered problems are all included to suggest to the pupil a little more definitely just what operations are present in the problems. In appearance, the conventional problems in Problem Test IV are like the imaginative; actually, they differ in the purpose for which the extra wording is used in each.

Problem Test V

It will be recalled that the element of time in Problem Tests I, II, and IV is only theoretically a constant factor. In these tests the pupils were given all the time they needed to try all the problems which were offered. In order to provide a check test in which the element of time would be held a constant, Problem Test V was designed.

Problem Test V is composed of two parts which are designated Problem Test A and Problem Test B. The two parts of the test are reproduced.

The directions for giving each part of Problem Test V are exactly the same as the directions for giving Problem Test II with the exceptions (1) that exactly ten minutes were to be taken for each part, and (2) the pupils were asked to try as many problems as they could in the time they had.

In comparing the two parts of Problem Test V, it will be seen that all factors are held constant save the manner of statement of the problems. The problem situations are similar, the operations involved are exactly the same, the difficulty is constant, and the testing time does not vary.

Problem Test A

Work as many of the following problems as you have time for; work them in order as numbered.

1. Tom was paid $1.20 an evening for working down town, but had to spend 12 cents each evening for car fare. After working 20 evenings last month how much did he have?

2. Mr. Brown sold celery at 9 cents a bunch. He found that if he trimmed it nicely he could get 12 cents a bunch. Last week he sold 87 bunches nicely trimmed. How much extra did he make by trimming it?

3. Two boys had to dig 43 rows of potatoes. They were paid 15 cents a row. Jim dug 23 rows and Henry dug the rest. How much did Henry make?

4. Tickets to the school fair were 30 cents each. In the afternoon 78 people came, and at night 228 people came. How much more money was taken in at night than in the afternoon?

5. Jack has to buy 25 gallons of gasoline each month for his car. Last month he bought it at 18 cents a gallon, but this month it costs 21 cents a gallon. How much more does he have to pay this month for gasoline?

6. For 20 days last month Mary's father gave her 35 cents a day for lunch money, but she got her lunch each day for 27 cents and saved the rest. How much did she save last month?

7. Henry is setting out plants in rows in his garden. In one end of each row he is setting out 4 tomato plants and in the other end 9 cabbage plants.

He will have 7 rows. How many more cabbage plants will he have than tomato plants?

8. The bookstore man bought 75 second-hand books for 60 cents each and sold them at 95 cents each. How much did he make?

9. The fruit man bought 25 dozens of oranges at 35 cents a dozen and sold them at 60 cents a dozen. How much did he make?

10. A grocer bought 30 dozens of eggs at 37 cents a dozen and sold them at 43 cents a dozen. How much did he make?

Problem Test B

Work as many of the following problems as you have time for; work them in order as numbered.

1. Sam lives in the country on a farm. He helps his father all he can, and his father always pays him. Last fall Sam helped pick up the potatoes when they were dug; he helped to put them in sacks to take to town to sell, and he helped to sell them. He sold 46 bushels at 85 cents a bushel. Sam's father said that all he needed to give him was 75 cents for each bushel and Sam could keep the rest. How much did Sam get to keep?

2. Jim is a pretty good gardener. He raised some nice head lettuce last fall. There was plenty for the family and Jim gave some to the neighbors and decided to sell the rest. He had 25 heads to sell. Mr. Jones, the grocer, said he would pay Jim 9 cents a head. Jim's father showed Jim how to trim it up nicely and told him to take it from door to door and try to sell it that way. Since that would take extra time and work, Jim decided to charge 15 cents a head. He sold all he had. How much extra did he make by selling his lettuce that way?

3. Mr. Smith had 38 rows of beans to be picked, and they all had to be picked one afternoon as he wanted to sell them the next day. George Johnson said he would pick the beans for 10 cents a row. That was all right, and so George got his smaller brother, Henry, to help him. Henry worked as hard as he could, but, of course, he could not pick beans as fast as George. They got the beans picked by sundown, and then they counted up and found that George had picked 25 rows and Henry the rest. When Mr. Smith paid the boys, how much should Henry get?

4. The boys and girls in the fifth and sixth grades at the Grant School were trying to raise some money. They wanted to buy books for their library, and they wanted to give something to the Children's Hospital. They talked over many plans for raising money and decided they would have a School Fair. The two grades had a contest selling tickets. The tickets were 25 cents each. The contest was close. The fifth grade sold 120 tickets, and the sixth grade sold 132 tickets. How much more money did the sixth grade make than the fifth?

5. The grocer on our corner has a delivery truck which he has to keep going all day. It uses up a lot of gasoline. In fact, it takes about 45 gallons of gasoline to run it each month. This month the price of gasoline went up from 17 cents a gallon to 21 cents a gallon, and the grocer thought he had better figure up his expenses on gasoline and everything else and see if he ought not to charge more for his groceries. How much more will our grocer have to pay for gasoline this month than he paid last month?

6. The Junior High School wanted to have a School Play at the end of the year. Their teachers were very busy and so it was decided to ask Miss Brown to coach the play and help the pupils have a good one. A committee went to see Miss Brown about it. They told her they would have no money to pay her until the play was given. "That will be all right," said Miss Brown, "you can just pay me 15 cents for every ticket you sell and keep the rest." That was agreed upon. The pupils sold 250 tickets at 35 cents each. How much money did they have after they had paid Miss Brown?

7. Betty and Billy went on a visit to their uncle's in the country just when most of the fruit trees were in bloom. Their Uncle Bob had a lot of fruit trees in his orchards, row after row of them, and all in bloom. What a pretty sight they made! Betty and Billy liked nothing better than to go with Uncle Bob among the trees and to guess what kind of fruit this one would have and what kind that one would have. In one field they noticed that the bloom at one end was pink and at the other end the bloom was white. They counted 18 rows of trees, and then asked Uncle Bob about the different colored blooms. "That's easy," said Uncle Bob, "the pink trees are peach, and the white ones are cherry trees. In each row there are 12 peach trees and 7 cherry trees." How many more peach trees were there in this field than cherry trees?

8. There is a newsboy in our block who delivers papers every afternoon. He always folds the papers nicely and gets them at the people's front doors, so they don't blow away all over the front yards like some I have seen. Last month he decided that he needed some extra money and so he made arrangements with the news dealer down town to sell magazines. That brought him in the extra money he needed. He sold altogether 30 magazines at 25 cents each. Since he had to pay the news dealer 18 cents each for the magazines, how much money did he make selling them?

9. As the Camp Fire girls were going to camp out for two weeks in the summer they wanted to make some extra money to buy the things they would need. They knew that nearly everybody in town would want to help them as much as they could, so they decided to sell candy. They made arrangements with the candy store to furnish the candy. They had to pay 58 cents a box for it. They bought 200 boxes and sold them from house to house at 75 cents a box. How much money did they make for their camp?

10. The Boy Scouts of Troop 13 wanted to make some money to build a camp. They knew that every loyal supporter of the high school football team

would want to show his colors at the next big game, so they bought 600 pennants at 18 cents each. Just before the big game they sold all the pennants for 25 cents each. How much money did they make for their camp?

Summary

Five series of tests have been used in the present investigation. Their designations and general characteristics are as follows:

Problem Test I is intended to be a means of comparing arithmetic problems of the conventional type with arithmetic problems of the imaginative type. The test contains ten pairs of problems. One member of each pair is a conventional problem, and the other an imaginative one.

Problem Test II is a repetition of Myers test comparing the two types of problems. It contains five pairs of problems.

The Collateral Tests are the Burgess Scale for Measuring Ability in Silent Reading, the Otis Self-Administering Test of Mental Ability, Intermediate Examination, selected parts of the Cleveland Survey Arithmetic Tests, Form 1, and the Stone Reasoning Test, Form 1. The Otis Test was used not merely as a measure of intelligence, but also as a measure of carefulness of work.

Problem Test IV is exactly a repetition of Problem Test I with the exception that the short conventional problems of Test I have been expanded in order to explain better the purely mathematical relations required in their solution.

Problem Test V is composed of two parts. Problem Test A is composed of ten short conventional problems and Problem Test B is composed of ten longer imaginative problems.

In Problem Tests I, II, and IV, the time element is a factor only in the sense that ample time was given every pupil tested in each to complete his trials of all the problems. In the Collateral Tests and in Problem Test V, the time element is taken into account and held as a constant factor.

CHAPTER IV

DIFFERENCES IN PUPIL-RESPONSES TO THE SHORTER CONVENTIONAL AND IMAGINATIVE PROBLEMS

This chapter will report the responses of 433 pupils in the fifth and sixth grades to the problems presented in Problem Test I and of 660 pupils in Grades 5 to 8 to the problems presented in Problem Test II.

Results of Problem Test I

During the last week in December, 1927, Problem Test I was given to 200 children in Grade 5B and 103 children in Grade 6B in the Huntington, West Virginia, Public Schools, and during the latter part of March, 1928, the test was given to children in the schools of Dayton, Ohio, and of Paducah, Kentucky, as follows: 31 pupils of Grade 5A in the training school of the Grace Green Normal School of Dayton, Ohio, and 36 pupils of Grade 5B, 34 pupils of Grade 5A, and 28 pupils of Grade 6B of the schools of Paducah, Kentucky. The results of the test in number of problems solved, that is, number of problems in which both operations were recognized, no account being taken of inaccuracies in computation, are given in Table 2.

Table 2 shows for each grade group and by totals a comparison of correct recognitions in the case of the members of the ten pairs of problems. In the first column the type problems are indicated, e.g., AA, Addition-Addition; AS, Addition-Subtraction, and so on, and immediately to the right the corresponding problem numbers. In the remaining columns appear the number of problems solved in each case, and for each grade group and for the total number of pupils tested the per cent of solutions. Thus, the total may be read, 104 of the 220 pupils of the Huntington 5B grades solved conventional problem 9, and 110 solved the corresponding imaginative problem 14; and so on. In the columns at the extreme right it is shown that of the 433 pupils tested, 297, or 69 per cent, solved conventional problem 9, and 304, or 70 per cent, solved the

TABLE 2

COMPARISON OF CORRECT SOLUTIONS OF CONVENTIONAL AND IMAGINATIVE PROBLEMS IN PROBLEM TEST I

PROBLEMS	HUNTINGTON 5 B		PADUCAH 5 B		TOTAL 5 B				DAYTON 5 A		PADUCAH 5 A	
	200		36		236				32		34	
	C	I	C	I	C	I	% C	% I	C	I	C	I
AA 9–14...	104	110	28	26	132	136	56	58	28	27	29	30
AS 15–10...	83	95	20	25	103	120	44	51	26	29	28	28
AM 7–4 ...	58	42	18	14	76	56	32	23	22	24	24	23
AD 1–6 ...	35	53	11	13	46	66	19	28	21	27	12	23
SS 11–16...	51	47	16	14	67	61	28	26	25	27	24	24
SM 3–12...	18	14	10	4	28	18	12	8	12	9	17	12
SD 19–2 ...	52	41	13	12	65	53	28	22	26	24	23	19
MM 13–18...	48	73	14	17	62	90	26	38	21	27	24	25
MD 17–8 ...	51	44	19	14	70	58	30	24	28	27	23	24
DD 5–20...	48	43	12	11	60	54	26	22	23	20	18	19
Total	548	562	161	150	709	712	30	30	232	241	222	227
Grand Total ..	1,110		311		1,421		...		473		449	
Differences ...	14	..	..	11	3	..	..	..	9	..	5	..

TABLE 2—*Continued*

TOTAL 5 A				HUNTINGTON 6 B		PADUCAH 6 B		TOTAL 6 B				GRAND TOTAL 5TH AND 6TH			
66				103		28		131				433			
C	I	% C	% I	C	I	C	I	C	I	% C	% I	C	I	% C	% I
57	57	86	86	83	85	25	26	108	111	82	85	297	304	69	70
54	57	82	86	78	84	24	25	102	109	78	83	259	286	60	66
46	47	70	71	59	53	21	20	80	73	61	56	202	176	46	40
33	50	50	76	48	57	17	20	65	77	49	59	144	193	33	45
49	51	74	77	57	61	21	22	78	83	60	63	194	195	45	45
29	21	44	32	29	19	14	10	43	29	33	22	100	68	23	16
49	43	74	65	65	57	22	22	87	79	67	60	201	175	46	40
45	52	68	78	67	80	24	21	91	101	70	77	198	243	45	56
51	51	77	77	63	64	22	24	85	88	65	67	206	197	48	45
41	39	62	59	50	53	12	20	62	73	47	56	163	166	37	38
454	468	69	71	599	613	202	210	801	823	61	62	1,964	2,003	45	46
922		...		1,212		412		1,624		...		3,967		...	
14	..	2	..	14	..	8	..	22	..	1	..	39	..	1	..

corresponding imaginative problem 14. The results given in the lower extreme of the table may be read: the 200 pupils in the Huntington 5B grades solved 548 conventional problems and 562 imaginative problems, a grand total of 1,110 solutions, and a difference of 14 solutions showing to the disadvantage of the conventional problems. In the columns giving the totals, the percentages and the percentage differences are given.

The most striking feature of the table is presented by the figures at the lower right-hand corner. Here it is to be seen that out of a grand total of 3,967 solutions, there is a difference of 39 to the disadvantage of the conventional problems or a difference of approximately 1 per cent.

Explanation of the Differences

Reasons for the differences in responses to the members of the various pairs of problems may be adduced by an analysis of their differences in method of presentation. To the members, respectively, of the Addition-Addition, Subtraction-Subtraction, and Division-Division problems, the differences in responses are negligible. The responses to the members, respectively, of the remaining pairs vary from slight to marked. The order of appearance of the quantities to be dealt with in the two members of a given pair of problems may be considered as a partial explanation of the difference in responses.

Table 3 indicates the order of appearance of the quantities in the various pairs of problems. The table is to be read as follows: In problems 9 and 14, the first and second quantities named are used in the first operation and are combined with their sum in the second operation as indicated; in problems 11 and 16, the first and second quantities named are used in the first operation, and the third quantity named is used with the result in the second operation, as indicated, and so on.

It will be seen that the order of appearance of the quantities in six pairs of the problems is the same in each case for the two members of the pair, while in four of the pairs the order is different. Comparing this observation with the results shown in the last columns of Table 2, one can find justification for the argument that the order of appearance of the quantities in a problem

TABLE 3

THE ORDER OF APPEARANCE OF THE QUANTITIES IN THE TEN PAIRS OF PROBLEMS

Kind of Problem	Example Number	Order of Appearance of the Two Required Operations: 1st Operation	Order of Appearance of the Two Required Operations: 2nd Operation	Comparison	Per Cent Correct Responses
AA	9	1, 2	(1 + 2), 1, 2	Same	69
	14	1, 2	(1 + 2), 1, 2		70
SS	11	1, 2	3, (1 — 2)	Same	45
	16	1, 2	3, (1 — 2)		45
DD	5	2, 1	3, (1 ÷ 2)	Essentially the same	37
	20	1, 2	3, (2 ÷ 1)		38
AS	15	2, 3	(2 + 3), 1	Different	60
	10	1, 2	(1 + 2), 3		66
AD	1	1, 2	3, (1 + 2)	Different	33
	6	2, 3	1, (2 + 3)		45
AM	7	1, 2	(1 + 2), 3	Different	46
	4	2, 3, 4	(2 + 3 + 4), 1 or 5		40
MD	17	1, 2	(1 × 2), 3	Different	48
	8	2, 3	(2 × 3), 1		45
SM	3	3, 2	(3 — 2), 1	Same	23
	12	3, 2	(3 — 2), 1		16
SD	19	1, 2	(1 — 2), 3	Same	46
	2	1, 2	(1 — 2), 3		40
MM	13	1, 2	(1 × 2), 3	Same	45
	18	1, 2	(1 × 2), 3		56

has something to do with the pupil's response to it. For example, the order is the same for the members of each of the AA, SS, and DD pairs, and to the two members, respectively, of each of these pairs the responses are about equal in number; whereas the order differs for the members of the AS, AD, AM, and MD pairs, and the number of responses in the case of each member of a given pair also is different. On the other hand, however, one notes the similarity in order of appearance of the quantities in the members of each of the remaining pairs, namely, the SM, SD, and MM problems and the differences in each case in the pupil's responses to the paired members. In explanation of differences in responses, one will have to look beyond mere similarity or difference in the order of appearance of the quantities presented, while at the same time he gives due weight to the order in which the quantities appear.

Passing over the AA, SS, and DD pairs, it will be noted in the AS pair that in problem 10 the quantities appear in the same order in which they have to be used, whereas in problem 15 the pupil is required either to keep the first quantity in mind until he is ready for it in the second operation or else to remember to return to it after the first operation has been performed. Here is a cause for incompleteness of work, because of the fact that the poorer pupils in arithmetic are inclined to be satisfied when they have done something with a problem and not to give due regard to the necessity of completing all that is required of them.

In the AD pair, it would appear that problem 1 is easier of solution than problem 6, because of the regular order of appearance of the quantities in the former. The greater number of pupils missing problem 1 may be explained by the fact that in the ordinary activities of life one is usually less efficient in the beginning, or "warming-up" process, of a task than he is immediately after.

In the AM pair, the reason for the fewer performances in problem 4 is readily apprehended. The multiplier 12 is mentioned twice. The double mention of 12, however, is part of the imaginative form of presentation.

In the MD pair, the advantage of regular order of appearance of the quantities is on the side of problem 17. On the other hand, the word "dozen" in problem 17 is a means of distraction in some cases, tempting some to multiply 45 by 12 before proceeding further.

Turning now to the three remaining pairs, the SM, SD, and MM problems, it may be pointed out that the disadvantage in statement or in type appears in each case to be against the conventional problem of the pair. In problem 3 of the SM pair, the word "dozen" is a distraction to some; in problem 19 of the SD pair the situation is not familiar to city children; and in problem 13 of the MM pair the inclination to reduce days to hours by using 24 or 12 as a multiplier is suggested to some.

In conclusion, it may be said, without regard to any scores actually obtained, that if any disadvantage exists in order of appearance of quantities or in statement or in type of situation presented that, on the whole, such disadvantage favors the imaginative rather than the conventional problems. In the AD, AM, and MD pairs, the advantage seems to be on the side of the conventional problems, but somewhat neutralized by the fact that the conventional problem of the AD pair is the first problem of the test. In the AS, SM, SD, and MM pairs, the advantage, if any exists, appears to be on the side of the imaginative problems. It is at least reasonably certain that the ten conventional problems of Problem Test I are not made easier or simpler, with respect to order of appearance of the quantities to be dealt with or terms used in the statement or computations demanded, than are the ten imaginative problems.

The Pupils Tested

In the administration of the test, the effort was made not to lose sight of the fact that the primary purpose was not to measure the abilities of the pupils tested but to compare their responses to two types of problems. In other words, a measure of the problems used in the test was desired in preference to a measure of pupil-ability. But since a measure of the problems used could be had only by inference and indirectly by means of the responses of pupils, consideration of the pupils tested is required.

In order to secure responses which will make comparison of the two types of problems possible, it is obvious that pupils must be selected for the test whose abilities to respond will be neither too low nor too high. For example, if the pupils' scores throughout are very low, very little comparative data can be secured. In the first place, many pupils would have to be tested in order to furnish a reasonable amount of responses for comparison, and, secondly, what is more important, the comparatively few correct responses which would be secured would be on the whole responses resulting from the accidental or chance solution of this problem or that one. On the other hand, if the abilities of the pupils tested are too high, the responses again would have little value for purposes of comparison. The very good pupil would solve all or nearly all of both types of problems and the differences in his responses to the two types of problems would be too small for purposes of comparison. Accordingly, the effort was made to select a group of pupils whose responses when averaged would score neither high nor low but approximately mid-way between the two extremes.

As has been indicated in the preceding chapter, the problems were selected by the writer and by members of his class from lists of problems actually in use in the fifth and sixth grades. Having decided upon two-step problems involving the four fundamental operations, the writer and the members of his class tried out various lists in informal tests in certain 4A and 5B grades. It quickly became apparent that too few of the 4A pupils could respond correctly to the two-step problems to justify giving Problem Test I in the 4A grades. Accordingly, the test was given first to 200 pupils in the 5B grades, as indicated in Table 2, with the result that 59 of the pupils tested failed to complete successfully the solution of a single problem. It was decided that if there was any particular virtue in either the conventional or imaginative type of problem, that virtue would hardly be discoverable in the responses of a group of whom one-fourth had missed all of the problems. Accordingly, the test was given later to other 5B pupils and to pupils in the 5A and 6B grades.

Table 2 shows the scores of the pupils in the 5A grades to be slightly higher than the scores of the pupils in the 6B grades. This is due to the fact that 6B scores were contributed largely by pupils in the Huntington schools. Since there is no purpose in the present investigation to establish grade norms, the slight disadvantage to be charged against the pupils in Huntington is not material.

Table 4 shows the distribution of scores for the pupils tested in the fifth grades and in the sixth grades and in both grades combined in terms of numbers of problems solved. It may be read: 20 pupils in the fifth grade solved 19 or 20 of the 20 problems of the test, 24 solved 17 or 18, 31 solved 15 or 16, and so on. The same data are presented graphically in Diagram I.

Diagram I shows a large number of pupils concentrated at the lower end of the scale, a large number at the upper end and fewer at each of the various points between. By no means can the distribution be said to approach that usually called the "normal," which is low at both extremes and higher and higher toward the center. The question arises: Does this lack of appearance of

TABLE 4

DISTRIBUTION OF PUPILS IN THE FIFTH AND SIXTH GRADES ACCORDING TO THE NUMBER OF PROBLEMS SOLVED IN PROBLEM TEST I

PROBLEMS	GRADES 5B AND 5A	GRADE 6B	TOTAL FIFTH AND SIXTH GRADES
	NUMBER SOLVING	NUMBER SOLVING	
19–20	20	19	39
17–18	24	21	45
15–16	31	18	49
13–14	17	17	34
11–12	21	13	34
9–10	21	8	29
7–8	20	11	31
5–6	15	13	28
3–4	26	6	32
1–2	48	6	54
0	57	1	58
Total	300	133	433

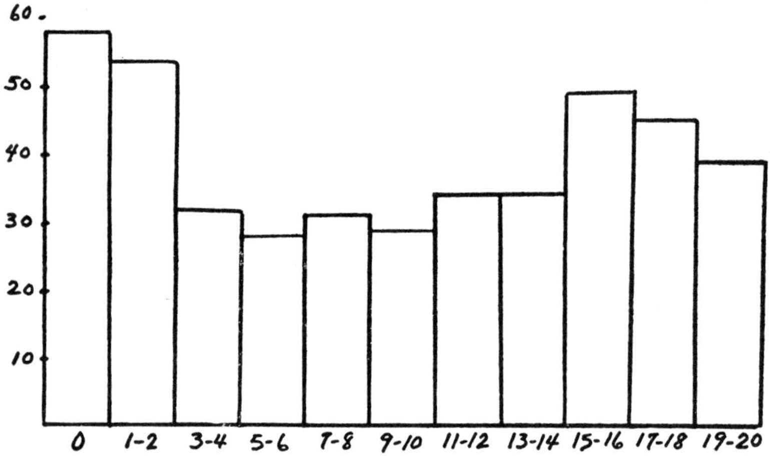

DIAGRAM I

DISTRIBUTION OF PUPILS IN THE FIFTH AND SIXTH GRADES ACCORDING TO THE NUMBER OF PROBLEMS SOLVED IN PROBLEM TEST I

The horizontal scale shows the number of problems solved; the vertical scale shows the number of pupils.

normality in the distribution of the scores in Problem Test I indicate that the problem-solving abilities of the pupils tested are also not normally distributed? In seeking an answer to the question, it is necessary to consider further the character of the test and the type of responses to be expected.

Problem Test I Not a Scale

The preceding chapter has indicated that Problem Test I is not a scale designed to measure the problem-solving abilities of pupils in the fifth and sixth grades. There it was indicated that the problem situations, the operations involved, and the difficulty of the parts are all theoretically constant, and that the test was so administered as to give each pupil all the time he needed to try all of the problems. The results presented in Table 2 indicate throughout that the twenty problems are very much the same in difficulty. The close approach to equality in difficulty is indicated still more impressively by Table 5.

TABLE 5

The Relative Difficulty of the Parts of Problem Test I

Problems	Per Cent Solving	Per Cent Failing	Percentile Value Scores
AA 9–14	69–70	31–30	42–41
AS 15–10	60–66	40–34	46–43
AM 7–4	46–40	54–60	52–54
AD 1–6	33–45	67–55	57–52
SS 11–16	45–45	55–55	52–52
SM 3–12	23–16	77–84	62–66
SD 19–2	46–40	54–60	52–54
MM 13–18	45–56	55–44	52–48
MD 17–8	48–45	52–55	51–52
DD 5–20	37–38	63–62	55–55

The data in the first and second columns of Table 5 are copied from Table 2 and the data in the third column are computed from those presented in column two. The fourth column gives the varying degrees of difficulty of the twenty problems, that is, their relative weights or values, stated in percentile scores. Their method of derivation is based upon the assumption that scholastic abilities fit the normal probability curve and that the percentage of pupils who solve a given problem corresponds to the percentage of area under the curve from the zero point to the accorded value point on the base line. Based upon such assumption, the data in column four are correct for the 433 pupils tested to the same degree that the distribution of their abilities (that is, their *actual* abilities and not those which the results of Problem Test I *appear* to indicate) approaches that of the normal. Indeed, it may be pointed out, if the data in column four are relatively incorrect, that corrections made upon the basis of a pupil-ability distribution which may the more closely approximate the normal would bring the percentile scores of the problems more closely together than column four indicates. The reason is that in a normal distribution the abilities in major part are grouped about a mid-point and in major part do not vary widely. Pupils with a normal distribution of abilities are thus expected to produce scores which are less varying. The differences in the percentile scores given, therefore, are either correct as they stand or they are less, not more, than the amounts actually given.

Since it was the purpose in using Problem Test I to discover the differences, if any exist, in pupil-responses to the conventional and imaginative problems when every pupil was given full opportunity to try them all, it is to be expected that the totals of the results would not appear as a "normal" distribution even if the abilities of the pupils tested should fit exactly the normal probability curve. Some pupils of less ability can be expected, if given ample time, to accomplish the same results as the pupils of greater ability actually accomplished in less time. Moreover, the character of the items of the test—situations in which learned or partially learned operations appear to test recognition—influences a piling up of scores at the two extremes of the distribution.

Chapter II has pointed out that the problems of the course in arithmetic are merely practice exercises intended to give training in recognizing operations already learned, and that one, if given ample opportunity, either recognizes the operations which appear in the problem situation or fails to recognize them. One either succeeds completely or fails completely. The problem either gives practice in recognition or it does not. There is no middle ground making possible scores which are half right and half wrong. As a consequence, a piling up of scores in the middle of the distribution is prevented, and the tendency for them to spread somewhat evenly throughout the distribution is encouraged.

Furthermore, it may be pointed out, Problem Test I is even less a test of the abilities of the good and the poor pupils than it is a test of the abilities of average ability. The better pupils had only twenty problems to try, and either could have solved more problems in the additional time the slower ones were taking to try all of them or else they could have solved more difficult problems than the ones presented. On the other hand, the 58 zero scores at the lower end of the distribution by no means indicate zero arithmetical ability for the 58 pupils concerned. If they had had easier problems, the 58 scores, as well as some immediately above the 58, would have spread down the scale.

In summary, it may be said that the character of the test, providing as it does twenty problems of approximately equal difficulty and given so that every pupil used as much time as he needed to try them all, produced exactly the type of results one would expect to secure with a group of pupils whose scholastic abilities fit closely the normal probability curve, that is to say, a distribution bunched at the extremes and scattering in the middle is the normal expectation. To secure results from a test like Problem Test I whose distribution would fit the normal curve, one would have to select a sampling of pupils whose abilities are not normally distributed, but approximately the same. It may be concluded, therefore, that the 433 pupils tested are a legitimate sampling of public school pupils for the purpose of providing results for a comparative study of the merits of the conventional and imaginative problems of the test. The fact that the general standing of the pupils is somewhat lower than is reasonably to be expected in the fifth and sixth grades is not material.

Reasons for an Unscaled Test

There are two important reasons why a test designed for such purposes as were to be met by Problem Test I should be unscaled: (1) A scaled test would have introduced elements beyond the scope of the present study which would have remained unmeasurable and a source of confusion; and (2) the materials of Problem Test I which are necessarily unscaled approximate in character and purpose the practice problems actually in use in the school.

In the production of a scale for the measurement of scholastic ability, it is necessary to assemble a number of elements of varying degree of difficulty or to combine equally difficult elements in various ways. Thus, in extending the limits of Problem Test I to include elements in a scaled relationship, one could either introduce more difficult elements involving fractions, decimals, percentage, and denominate numbers, or he could introduce one-step problems in the fundamental operations which are less difficult and three- and four-step problems which are more difficult. Either undertaking would enlarge unreasonably the scope of the present study.

The previous chapter indicates the limits of the present study in so far as the operations involved are concerned, and suggests the reason for the limitation, which is that pupils in the fifth and sixth grades have all been instructed in the four fundamental operations, and will meet some little difficulty with them when they are combined in two-step problems. To state the reason in another way: The limitation guarantees problems easy enough to be solved and hard enough to present some difficulty. If one were to introduce elements like fractions, decimals, percentage, denominate numbers, and three- or four-step problems, he could no doubt produce a scale containing conventional problems and a scale containing imaginative ones. One would have no means of knowing, however, just what combinations of such elements would produce the relatively more difficult problems or steps in the scale in the case of either the conventional or imaginative half of the test, and, what is more to the point, he would have no assurance that similar combinations of elements in the two halves would result in corresponding steps of difficulty. It must be set down as a general principle that one cannot be certain of the results of a test if it contains many elements or elements in uncertain combinations of difficulty. One must narrow the scope of a study to be at all certain of his results.

Turning to the second reason why Problem Test I has been built into an unscaled test, one is again impressed with the fact that the problems of the arithmetic class in the school are not primarily test exercises, but practice exercises. The pupil first learns through instruction or study or both a principle or procedure; and, secondly, he fixes or generalizes it through practice in applying it in various situations. For example, the pupil moves from the mastery of the one-step problem to a consideration of problems with two-steps. The idea of a double operation develops gradually. Its development is hastened through special practice in giving consideration to situations which involve a two-step series of operations. The problem of the present investigation—to compare the merits of conventional and imaginative problems as exercises in the generalization of knowledge of the fundamental operations—can be attacked systematically only by making the problems of the testing exercises similar to the problems of the practice exercises of the school.

The Point Score

In order to provide measures which are comparable in the case of the poorer pupils as well as in the case of the better ones, the number of points secured by each pupil was taken as a measure of his responses to the problems. The point score—two points for each problem solved, and one point for each problem partly solved—was determined in the case of each pupil, first, for the ten conventional problems and, next, for the ten imaginative problems, in the manner indicated in the foregoing chapter. The pupil's score, representing the superiority of problems of the one type over problems of the other type, was computed by taking the difference between his two-point scores. Thus, in each case the number of points secured in the ten imaginative problems was subtracted from the number of points secured in the ten conventional ones, and the difference was set down as the pupil's C-I point score. Thus, a point score with the positive sign shows a superiority in favor of the conventional group, and a point score with the negative sign shows a superiority in favor of the imaginative group.

Distribution of the Point Score

The distributions of the point scores for Grades 5 and 6 are shown in Tables 6 and 7. Table 7 differs from Table 6 in the respect that it does not include the pupils who failed to complete any of the problems in Problem Test I. Table 6 may be read as follows: In the fifth grades, 6 pupils were more successful in the imaginative problems by 5 points, 10 pupils made a higher score by 4 points, 11 by 3 points, and so on. Table 7 may be read the same as Table 6. The figures to the right of the zero column indicate the number of pupils who were more successful in the conventional problems. In the columns to the right are given the totals of the pupils and their average point scores by grades.

TABLE 6

DISTRIBUTIONS OF C–I POINT SCORES IN PROBLEM TEST I

C–I POINT SCORES	−5	−4	−3	−2	−1	0	1	2	3	4	5	TOTAL	AVERAGE SCORE
Fifth Grades ..	6	10	11	30	46	86	55	22	17	8	9	300	.06
Sixth Grades ..	5	6	10	16	22	31	19	12	4	5	3	133	—.31
Total	11	16	21	46	68	117	74	34	21	13	12	433	—.06

TABLE 7

DISTRIBUTIONS OF C–I POINT SCORES IN PROBLEM TEST I
(Pupils with zero problem scores are not counted)

C–I POINT SCORES	−5	−4	−3	−2	−1	0	1	2	3	4	5	TOTAL	AVERAGE SCORE
Fifth Grades ..	6	9	10	27	40	58	39	20	17	8	9	243	.07
Sixth Grades ..	5	6	10	16	22	31	18	12	4	5	3	132	—.33
Total	11	15	20	43	62	89	57	32	21	13	12	375	—.07

The most impressive fact presented in each of Tables 6 and 7 is the overwhelming number of pupils crowding the distribution at or about the zero score. In every instance approximately one-fourth of the scores recorded show no difference in pupil-responses to the two types of problems, and another one-fourth and more show a difference one way or the other of only one point. When it is considered that a difference of one point in any case is so small as to justify its being regarded as due to chance and hence of zero value, it can be seen that the tables offer a preponderance of evidence that for the majority of pupils the one type of problem practice exercise is just as good as the other. Indeed, there is such a regular and gradual dropping off of the measures to the right and left of the zero score as to convince one that a point score showing a difference greater than one is probably due to chance and is representative of a difference that is more accidental than real. Moreover, when in any case all the scores are combined into an average score, neither type of problem shows to advantage in comparison with the other. One is justified in the conclusion that when the practice period of pupils is not limited as to time, they are not hindered in their practice exercises by using problems of the longer, imaginative type, and, what is more to the point, they are not helped.

Explanation of the Distributions

On the other hand, it may be pointed out that the significance of the distributions which are shown in Tables 6 and 7 should be sought in the variations of the scores from the central tendency, or average, quite as much as in the central tendency itself. It may be suggested that the scores at the extremes of the distributions are the significant scores. Perhaps the scores at one extreme are those of pupils who have great difficulty in their problem practice exercises, while the scores at the other extreme belong to those who find little difficulty. In such a case the scores at the extremes would be scores of great significance.

It was decided during the progress of administering Problem Test I to secure enough data from the pupils taking the test to determine whether there was any indication that the differing characteristics of the pupils contributed to the differences in their scores. Should a marked indication be discovered, then the matter could be pursued by further testing; but if, on the other hand, little or no indication was discoverable, then the question could be left open for later study in connection with Problem Test IV which was considered capable of producing point scores farther removed from the zero than those of Problem Test I.

The Influence of Reading Ability

Since ability to read relates to ability to solve a printed arithmetic problem, and since the imaginative problems present more material to be read than the conventional problems do, the first comparison to suggest itself was that between the point scores and scores in reading. The Burgess Scale for Measuring Ability in Silent Reading was given to 160 pupils in the fifth grades. The results of the test in reading in comparison with the point scores for the same pupils are shown in Table 8.

TABLE 8

THE RELATION BETWEEN READING ABILITY AND THE C-I POINT SCORES IN PROBLEM TEST I

BURGESS READING SCORES	POINT SCORES											TOTAL
	−5	−4	−3	−2	−1	0	1	2	3	4	5	
13............					2	1		1				4
12............								1		1		2
11............		1		1	1	1	1		1			6
10............			2		1	1	1	1				6
9............				2	2	1	2	1	1			9
8............	1	2	1	2	4	3	2			1		16
7............	2	1		1	2	8	3	1	2		1	21
6............	1	1	1		2	3	1	1				10
5............		1	1	3	5	12	5	1	3			31
4............	1			4	3	7	7		4	2		28
3............			2			11	5	4	1	2	2	27
Total........	5	6	7	13	22	48	27	11	12	6	3	160

$r = -.16 \pm .051$

Table 8 shows a double distribution. From left to right the point scores are shown and from bottom to top the scores in reading are shown. Except along the column for the zero point scores from bottom to top, there is not seen to be exhibited any marked tendency for the scores to cluster at any particular point in the table; but, on the whole, they are scattered to all points throughout the table. There is a slight indication that the pupils who are very poor in reading do better in the conventional problems than in the imaginative. Several of the good readers, however, have corresponding point scores, and a few of the poor readers have point scores at the left extreme of the table. The coefficient of correlation between the two sets of measures was computed by means of the Pearson product-moment formula,

$$r = \frac{\Sigma xy}{N\sigma_x\sigma_y}$$

and the probable error of the coefficient was computed by means of the formula,

$$P.\,E.\text{ of } r = .6745\frac{1 - r^2}{\sqrt{N}}$$

The correlation thus determined, $r = -.16 \pm .051$, indicates a very low and a very unreliable relationship between low scores in reading and points in favor of the conventional problems and between high scores in reading and points in favor of the imaginative problems. The arrangement of the measures throughout the table indicates the unreliability of the computed relationship much more impressively than the coefficient itself. The arrangement indicates that the probability is high that a poor reader will be about equally unsuccessful in the two types of problems, and that a good reader will be about equally successful in the two types.

The Influence of Ability in Problem Exercises

If there is any marked relation between ability to solve problems in arithmetic and the tendency to react more favorably to one type than to the other, as represented by the point scores, that relation will be indicated in a comparison of the point scores with the problem scores. To be sure, the problem scores in Problem Test I are not measures of abilities. On the other hand, they provide some indication of the abilities of the pupils who produced them. Consequently, any clearly indicated degree of relation between the two sets of scores may be taken to justify further study, whereas a very slight indication of relationship may be taken as sufficiently conclusive that further study is unnecessary.

Table 9 shows the relation between the numbers of problems solved and the point scores for the pupils of the fifth grades, and Table 10 shows a similar relation for the pupils of both fifth and sixth grades. The coefficient of correlation computed from Table 9 is $r = -.02$; the coefficient of correlation computed from Table 10 is $r = -.04$. In either case, the zero row of scores was eliminated from the computations.

TABLE 9

THE RELATION BETWEEN THE TOTAL PROBLEM SCORES AND THE C–I POINT SCORES IN PROBLEM TEST I—FIFTH GRADES

Problem Scores	Point Scores											Total
	−5	−4	−3	−2	−1	0	1	2	3	4	5	
19–20					6	8	6					20
17–18		1	1	1	1	10	5	4		1		24
15–16		1	1	8	5	8	4	3		1		31
13–14		3	1	1	2	5	2	1	1	1		17
11–12				2	6	3	1	6	2		1	21
9–10	2	1		2	3	4	3	1	4		1	21
7–8			1	2	3	6	2	1	2	2	1	20
5–6	3	1		1	1	2	2		1	2	2	15
3–4	1	1	4	4	1	2	5	2	2	1	3	26
1–2		1	2	6	12	10	9	2	5		1	48
0		1	1	3	6	28	16	2				57
Total	6	10	11	30	46	86	55	22	17	8	9	300
Total (not counting zero row)	6	9	10	27	40	58	39	20	17	8	9	243

$r = -.02$

TABLE 10

THE RELATION BETWEEN THE TOTAL PROBLEM SCORES AND THE C–I POINT SCORES IN PROBLEM TEST I—FIFTH AND SIXTH GRADES

Problem Scores	The Point Scores											Total
	−5	−4	−3	−2	−1	0	1	2	3	4	5	
19–20				1	10	17	10	1				39
17–18		1	2	3	5	15	11	7		1		45
15–16		2	4	11	8	10	6	5	2	1		49
13–14	1	3	2	3	7	10	4	1	1	2		34
11–12	3	1	1	4	6	5	1	7	3	1	2	34
9–10	2	3	1	3	5	5	4	1	4		1	29
7–8			2	5	5	9	2	3	2	2	1	31
5–6	4	2	1	1	1	4	5	2	1	5	2	28
3–4	1	2	5	4	2	3	5	2	3	1	4	32
1–2		1	2	8	13	11	9	3	5		2	54
0		1	1	3	6	28	17	2				58
Total	11	16	21	46	68	117	74	34	21	13	12	433
Total (not counting zero row)	11	15	20	43	62	89	57	32	21	13	12	375

$r = -.04$

Results of Problem Test II

During the early weeks of February, 1928, Problem Test II was given to 660 pupils in Grades 5B to 8A in the Huntington Public Schools. The results in total numbers of problems solved are shown in Tables 11 and 12. Table 11 indicates that 21 pupils in the 5B grade took the test and of this number 5 solved problem 1, 11 solved problem 2, 0 solved problem 3, and so on. The bottom row shows the totals for the 660 pupils. In Table 12 the totals are grouped in paired arrangement to show the comparisons in correct solutions between the two types of problems. Thus, it is shown that 402 pupils solved conventional problem 8, and 370 pupils solved imaginative problem 1, its paired opposite, and, in like manner, problem 7 is compared with problem 2, problem 3 with problem 5, problem 4 with problem 9, and problem 6 with problem 10.

TABLE 11

The Results of Problem Test II in Terms of Correct Responses to the Various Problems

Grade	Number Pupils Tested	Correct Solutions of Problems as Numbered									
		1	2	3	4	5	6	7	8	9	10
5B	21	5	11	0	2	9	3	10	9	2	2
6B	66	9	22	15	1	25	15	21	19	5	20
6A	81	30	40	37	13	44	32	49	44	15	37
Total 6th	147	39	62	52	14	69	47	70	63	20	57
7B	151	77	112	60	32	92	84	117	82	46	69
7A	91	56	68	34	26	63	57	66	56	32	53
Total 7th	242	133	180	94	58	155	141	183	138	78	122
8B	160	121	139	92	60	136	124	144	125	86	122
8A	90	72	76	50	35	75	73	86	67	45	72
Total 8th	250	193	215	142	95	211	197	230	192	131	194
Total 5B-8A	660	370	468	288	169	444	388	493	402	231	375

In three of the five pairs of problems, namely, the 9-1, 7-2, and 6-10 pairs, the scores indicate that the short, tersely-stated,

TABLE 12

Comparison of Correct Solutions of Conventional and Imaginative Problems in Problem Test II—660 Pupils

Conventional		Imaginative	
Problem Number	Times Solved	Problem Number	Times Solved
8	402	1	370
7	493	2	468
3	288	5	444
4	169	9	231
6	388	10	375

conventional problems are the more successful in calling forth correct responses than are the problems of the longer, wordier, imaginative type. In the other two pairs, exactly the opposite indication is to be noted in the results given.

Explanation of the Results

When one examines the statements of the problems in Problem Test II, which is reproduced in the preceding chapter, he is able to discover certain indications of the reasons for the differences in the scores which are presented in Table 12. For example, one can find no apparent difference between problems 8 and 1, or between problems 7 and 2, or between problems 6 and 10, except the intended difference in manner of statement, which is the element being tested. For the members of each pair, the operations involved are about equal in degree of difficulty, and so are the situations and the manner of describing the situations. There is no apparent reason why the pupils respond slightly more successfully to the three conventional problems of these pairs unless it is that the wordiness of the imaginative problems is a hindrance and a means of distraction.

Turning to the two remaining pairs, namely, the 3-5 and the 4-9 pairs, one can discover in the manner of presenting the situations certain clues leading to an explanation of the more successful responses to the imaginative members. For example, in problem 3 the manner of describing the situation fails to make it entirely clear to pupils whether Laura's grandmother gave her $1.50 a day, or "had given" her $1.50 and no more. The careless formulation of the problem, resulting in the statement of the two clauses of the first sentence in the same tense, operates as a source of confusion. The teachers who assisted the writer in giving the test reported a number of pupils who raised questions as to the exact meaning of the statement of problem 3.

Likewise, problem 4 is a very poor statement of the situation being presented as compared with problem 9. Problem 4 is stated in a single sentence, which contains four separate statements of quantitative relations, and includes a question at the end. There is no breathing place provided at any point within the problem where the reader can collect his thoughts before reading ahead. Its manner of presenting the problem situation discourages careful deliberation as one proceeds with the reading of it. One is forced to conclude that pupils might well be expected to have difficulty with problems 3 and 4 regardless of whether they are stated in the conventional form or in the imaginative form.

The Results of Myers' Study

The preceding chapter has pointed out that Problem Test II is a partial repetition of a similar test given by Garry C. Myers, of the Cleveland School of Education.[1] The results which Myers has reported arc reproduced in Tables 13 and 14. Tables 13 and 14 are to be read in the same manner as Tables 11 and 12 with which they are to be compared. Myers' results appear to indicate a pronounced superiority in favor of the imaginative problems, and he concludes so emphatically.

In studying Myers' results, as they appear in Table 13, one cannot fail to notice the corresponding scores of the fifth- and sixth-grade pupils in conventional problems 3 and 7. One cannot avoid asking the question: If 152 pupils in the fifth grade were good enough in arithmetic to produce a score of successful solutions of problem 3, amounting to 59, and a score of successful solutions of problem 7, amounting to 93, how can it be explained that 272 pupils in a grade one year in advance of the fifth were able to

[1] Garry C. Myers, *Errors in Arithmetic,* pp. 65-68. The Plymouth Press, Chicago, 1925.

TABLE 13

THE RESULTS OF PROBLEM TEST II AS REPORTED BY MYERS

Grade	Number Pupils Tested	Correct Solutions of Problems as Numbered									
		1	2	3	4	5	6	7	8	9	10
5	152	70	97	59	16	76	48	93	79	25	59
6	272	160	201	59	46	163	88	93	154	64	125
7	37	25	36	26	13	33	22	32	24	19	26
8	33	29	31	26	5	29	23	28	30	12	26
Normal Students ...	19	18	19	11	11	19	16	17	15	14	17
Total	513	302	384	181	91	320	197	263	302	134	253
Total (omitting sixth grade scores)	241	142	183	122	45	157	109	170	148	70	128

do no better? The lower row in Table 13 shows the results reported by Myers with his sixth-grade scores omitted, and Table 15 is the same as Table 14 with the sixth-grade scores omitted. One cannot help but think that perhaps the results which are shown in Tables 13 and 14 have been subjected to the same carelessness in reporting as is exhibited in the manner of stating conventional problems 3 and 4. There certainly is little ground in the results for the somewhat extravagant conclusion that "there can be no question about the greater interest which such problems [the imaginative] stimulate. It is difficult to get fifth- and sixth-grade

TABLE 14

COMPARISON OF CORRECT SOLUTIONS OF CONVENTIONAL AND IMAGINATIVE PROBLEMS, AS REPORTED BY MYERS—513 PUPILS

Conventional		Imaginative	
Problem Number	Times Solved	Problem Number	Times Solved
8.....	302	1.....	302
7.....	263	2.....	384
3.....	181	5.....	320
4.....	91	9.....	134
6.....	197	10.....	253

TABLE 15

COMPARISON OF CORRECT SOLUTIONS OF CONVENTIONAL AND IMAGINATIVE PROBLEMS, AS REPORTED BY MYERS, WITH SCORES OF SIXTH-GRADE PUPILS OMITTED—241 PUPILS

Conventional		Imaginative	
Problem Number	Times Solved	Problem Number	Times Solved
8.....	148	1.....	142
7.....	170	2.....	183
3.....	122	5.....	157
4.....	45	9.....	70
6.....	109	10.....	128

children away from them."[2] Indeed, a comparison with the results of Problem Test II compels the conclusion that the imaginative type of statement of a problem is not the superior of the conventional type, and a critical analysis of the sentence structure of certain of the conventional problems does not detract from the weight of the evidence which supports it.

Summary

1. The results of Problem Test I show negligible differences between the totals of responses to the two types of problems. Slight differences in difficulty between the members of certain pairs of the problems are shown.

2. The order of appearance of the quantities in the statement of certain problems, and suggestions to use one's knowledge of denominate numbers, for example, "12 in a dozen," and "24 hours in a day," when the use of such knowledge is not demanded, are factors which account for slight differences in difficulty.

3. On the whole, the two sets of ten problems each are about equal in degree of difficulty. Any disadvantage due to differences in difficulty between the paired members appears to favor the imaginative problems.

4. Although the results when expressed as problem scores do not show up as a "normal" distribution, the fact that Problem Test I is not a scale intended to measure ability, coupled with the "normal" distribution of the point scores, indicates that the abilities of the pupils tested are fairly normally distributed and constitute a legitimate sampling of intermediate grade pupils.

5. There appears to exist no reliable measure of relation of the point scores to reading ability or to problem-solving ability as shown by the results in problem scores. The distribution of the point scores seems to be due almost entirely to chance.

6. The results of Problem Test II indicate a superiority of the conventional problems over the imaginative ones, when the carelessness of the statements of two of the conventional problems is taken into account.

7. The results of Myers' study present no convincing evidence of the superiority of imaginative problems.

[2] Garry C. Myers, *Errors in Arithmetic*, p. 68.

In brief, pupils of the intermediate grades are neither hindered nor helped in their problem practice exercises by problems of the imaginative type, when no limits are imposed upon the amounts of time of their practice periods.

CHAPTER V

DIFFERENCES IN PUPIL-RESPONSES TO THE LONGER CONVENTIONAL AND IMAGINATIVE PROBLEMS

This chapter will report the responses of 631 pupils in the fifth and sixth grades to the problems presented in Problem Test IV, and for 293 pupils in the sixth grade will present comparisons of these responses with the scores in the Collateral Tests.

THE REASONS FOR PROBLEM TEST IV

The outstanding conclusion to be drawn from the results reported in the foregoing chapter—that the imaginative type of practice problem in arithmetic presents no element of superiority over the conventional type of problem—suggests the possibility of discovering some element, either of superiority or inferiority, in a modification of the problems of the conventional type.

In Problem Test I special pains were taken to expand the even-numbered problems to include certain possible imaginative elements of the situations which these problem statements presented. In view of the results which were secured, one may be led to inquire what would be the effect if the same degree of extra effort is expended in enlarging the odd-numbered problems to include elements which are designed to render the situations presented in the conventional statements more easily recognizable. Since no essential differences in pupil-responses were discoverable in the comparative study of the "shorter" conventional with the imaginative problems, it was decided to expand Problem Test I into Problem Test IV with the view to providing, if possible, some measurable difference in the responses to the two types of problems. The expansion undertaken was in the direction of making the situations in the odd-numbered, or conventional, problems less difficult to understand by describing more fully and indicating more specifically what may be termed their strictly mathematical elements.

The Pupils Tested

During the early weeks of February, 1928, Problem Test IV was given to 293 children in the 6B grades in the Huntington, West Virginia, Public Schools. Reference to Table 2 will indicate the reasons for selecting the 6B grades. Table 2 shows the December, 1927, results of Problem Test I in the 5B and 6B grades. The inference to be drawn from these results is that for the purposes of the present investigation a problem test of the difficulty of Problem Test I is somewhat too difficult for the Huntington pupils of the 5B grades. It was decided therefore to give Problem Test IV in Huntington to the pupils of the 5A grades. Since the test was given immediately after the school mid-year when the 5A pupils had moved along to the 6B grades, the pupils tested are designated as 6B pupils.

In order further to compare the two types of problems, the test was given to pupils in corresponding grades in the schools of Racine and Sheboygan, Wisconsin. Through the coöperation of Mr. F. S. McCall, principal of the Knapp School in Racine, Wisconsin, and of Mr. C. F. Lutze and Mr. E. H. Weisler, elementary principals and members of the teachers committee on arithmetic in Sheboygan, Wisconsin, the test was given in these cities under conditions identical with those which prevailed in the administration of the test in Huntington. The test was given during the early weeks of March, 1928, to 254 pupils in the fifth grades in Sheboygan and to 84 pupils in the fifth grades in Racine.

Results of Problem Test IV

The results of the test in number of problems solved, that is, number of problems in which both operations were recognized, no account being taken of inaccuracies, are given in Table 16.

Table 16 shows for each city and by totals a comparison of correct recognitions in the case of the members of the ten pairs of problems. In the first column the type problems are indicated, thus, AA, Addition-Addition, AS, Addition-Subtraction, and so on, and immediately to the right the corresponding problem numbers. In the remaining columns appear the number of problems solved in each case, and for each city and for the total number of pupils tested

TABLE 16

Comparison of Correct Solutions of Conventional and Imaginative Problems in Problem Test IV

Problems	Huntington 6B				Sheboygan 5th		Racine 5th		Total (Other Cities) 5th				Total 5th and 6th			
	293				254		84		338				631			
	C	I	% C	% I	C	I	C	I	C	I	% C	% I	C	I	% C	% I
A A 9–14.......	180	203	62	69	152	216	58	68	210	284	62	84	290	487	62	77
A S 15–10.......	204	206	69	70	210	224	62	68	272	292	80	86	476	498	76	79
A M 7–4	181	144	62	49	215	153	50	42	265	195	78	58	446	339	70	54
A D 1–6	128	141	44	48	188	183	32	39	220	222	65	65	348	363	55	58
S S 11–16.......	168	164	57	56	209	169	51	37	260	206	77	62	428	370	68	59
S M 3–12.......	67	44	23	15	92	65	14	8	106	73	30	21	173	117	26	19
S D 19–2	172	124	58	42	202	191	37	32	239	223	71	65	411	347	65	55
M M 13–18.......	138	179	47	61	118	172	19	48	137	220	42	65	275	399	44	64
M D 17–8	161	140	55	48	186	184	37	30	223	214	65	63	284	254	61	56
D D 5–20.......	121	116	41	40	147	182	35	38	182	220	54	65	303	336	48	53
Total	1,520	1,461	52	50	1,719	1,739	395	410	2,114	2,149	63	64	3,634	3,610	58	57
Grand Total ...	2,981		....		3,458		805		4,263		....		7,244		....	
Difference		59		2	20		15		35		1			24		1

the per cent of solutions. Thus, the table may be read, 180, or 62 per cent, of the 293 pupils tested in Huntington solved conventional problem 9, and 203, or 69 per cent, solved the corresponding imaginative problem 14, and so on. In the columns at the extreme right, it is shown that 290, or 62 per cent, of all the pupils tested, solved, conventional problem 9, and 487, or 77 per cent, solved imaginative problem 14, and so on. In the lower rows of the table, totals of solutions of the two types of problems are compared with each other and with their sums and differences. Thus, in the columns showing the scores of the Huntington pupils, it is indicated that the 293 pupils made a total score of 1,520 in the conventional problems, and of 1,461 in the imaginative, resulting in a grand total of correct solutions of 2,981 and in a difference to the disadvantage of the imaginative problems of 59. Percentages and differences in percentages are shown in a similar manner.

Differences in Responses

The totals given in the lower rows of Table 16 show slight differences in the responses of pupils to the two types of problems. For the Huntington pupils, there is a slight difference in favor of the longer conventional problems; for the pupils in the two other cities, the difference which exists favors the problems of the imaginative type. On the whole, it would appear that there is little, if anything, to be gained by attempting to enlarge the practice effects of problems of the conventional type through extra wording which is designed to explain more clearly the purely mathematical elements of the problem situation. Comparing the totals given in Table 16 with the corresponding totals in Table 2 which show the results of Problem Test I, one is influenced toward the conclusion that problems of the short, tersely-stated, conventional type might just as well be left with their statements undisturbed. On the other hand, an examination of individual pairs would indicate the possibility in certain instances of improving a statement of a problem by enlarging it.

Influences of the Longer Conventional Statements

The matter may be approached through a comparison of the results in the two tests as they are given in the columns showing the totals in Tables 2 and 16. The data for such a comparison have been assembled in Table 17. The facts of major significance are those to be found in the columns which show the percentage comparisons. Thus the table may be read: In Problem Test I, the number of responses to short conventional problem 9 was 99 per cent of the number of responses to imaginative problem 14 and, in Problem Test IV, the corresponding percentage was 81. Thus the attempt to improve problem 9 through extra wording failed to the extent of a decrease of 18 in the percentage of conventional to imaginative responses. Failure is likewise shown in the comparisons of the SM, MM, and DD problems. The attempt to improve the statements of the short conventional problems in the AS, AM, AD, and SS pairs resulted in a marked increase in percentage of correct responses from Test I to Test IV, and in the SD and MD pairs only a slight increase is shown.

TABLE 17

COMPARISON OF THE TOTALS OF CORRECT RESPONSES TO PROBLEM TESTS I AND IV

PROBLEMS		PROBLEM TEST I					PROBLEM TEST IV					
		NUMBER TESTED—433					NUMBER TESTED—631					
		C	I	% C	% I	% C OF I	C	I	% C	% I	% C OF I	INCREASE
AA	9–14	297	304	69	70	99	390	487	62	77	81	−18
AS	15–10	259	286	60	66	91	476	498	76	79	96	5
AM	7–4	202	176	46	40	115	446	339	70	54	130	15
AD	1–6	144	193	33	45	73	348	363	55	58	95	22
SS	11–16	194	195	45	45	100	428	370	68	59	115	15
SM	3–12	100	68	23	16	144	173	117	26	19	137	− 7
SD	19–2	201	175	46	40	115	411	347	65	55	118	3
MM	13–18	198	243	45	56	80	275	399	44	64	69	−11
MD	17–8	206	197	48	45	107	384	354	61	56	109	2
DD	5–20	163	166	37	38	97	303	336	48	53	91	− 6

The statements of problems 19 and 17 might just as well have been left as they were in Problem Test I, while the lengthened statements of problems 9, 3, 13, and 5 seem actually to have been more a source of confusion than of help. The lengthened statement of problem 9 seems to have been especially confusing to the pupils in the cities of Sheboygan and Racine. In his comment upon the testing in his school, Principal McCall of Racine writes: "The children found the question in problem 9 not clear. Some thought individual amounts were asked for, while others thought a collective amount was asked for. Some tried to find one answer, some three." Thus we have a case of an easy problem made difficult through an attempt to explain the obvious. The lengthened statement of problem 3 seems also to have been a hindrance. The results of Problem Test I show problem 3 to be very difficult. It remains a difficult problem in spite of the effort to explain the mathematical elements of the situation presented in its statement. Problem 5 is comparable to problem 3, whereas problem 13 presents a somewhat different illustration. The situation presented in problem 13 seems much more difficult for the pupils to grasp than does the situation in problem 18. Its lengthened statement increases its difficulty. The foregoing comparison may be taken to indicate the danger of contributing to the pupil's confusion which may result from the effort to aid his recognition of the operations involved in a problem. When the problem is easy, the pupil does not need a lengthy explanation; when it is difficult, he may need a type of assistance which a lengthy explanation does not give.

Further reference to Table 17 would indicate the advisability of excluding the results of the AA, SM, MM, and DD problems from the final comparisons. It is one of the purposes of this chapter to study the reasons for improvement when improvement results from a lengthening of the statements of the odd-numbered problems. Consequently, it would seem advisable in meeting this particular purpose to exclude those results which have been influenced by unsuccessful attempts to render the shorter conventional problems more understandable. Moreover, the results of the AA pair ought to be excluded on the ground that problem 9 in Test IV is more an imaginative problem of the usual type than it is a conventional problem of the longer type.

The fact that a legitimate comparison will result from the exclusion of the results of the four pairs of problems is shown in Table 18. The table shows, both for the Huntington pupils and for all of the pupils, only slight differences in the responses to the paired members of the six pairs of problems in Problem Test I, and marked differences in the case of Problem Test IV. The table seems to indicate that the comparisons sought in Problem Test IV will not be prejudiced through any pronounced differences in the results of Problem Test I that may be brought about by the exclusion of the four pairs of problems.

TABLE 18

COMPARISON OF THE TOTALS OF CORRECT RESPONSES TO PROBLEM TEST I AND PROBLEM TEST IV WHEN THE AA, SM, MM, AND DD PAIRS ARE EXCLUDED

Test	TEST I				TEST IV				TEST I				TEST IV			
Pupils	HUNTINGTON—303				HUNTINGTON—293				TOTAL—433				TOTAL—631			
Problems ..	C	I	% C	% I	C	I	% C	% I	C	I	% C	% I	C	I	% C	% I
Total Responses ..	700	688	38	38	1,014	919	58	52	1,206	1,222	46	47	2,493	2,271	66	60
Grand Totals ...	1,388		...		1,933		...		2,428		...		4,764		...	
Differences .		12		0		95		6	16		1			222		6

FREQUENCIES OF UNPAIRED SOLUTIONS

The influence of the longer conventional statements may be shown in another way. Table 19 shows a comparison of the frequencies and their percentages of unpaired solutions in Problem Test I with the corresponding frequencies and percentages in Problem Test IV. A method of studying the relative ease of solution of a given pair of problems is that of determining the number of times the one was solved by a given set of pupils when the other remained unsolved, and vice versa, and of computing the percentage of the times each was solved alone to its total solution. Table 19 presents the results of such a method.

Table 19 may be read as follows: Problem 9 of the AA pair of Problem Test I was solved by 27 pupils who failed to solve problem 14. This is 9 per cent of the total of the pupils who solved problem 9. Problem 14 was solved by 34 pupils who failed to solve problem 9. This is 11 per cent of the total of pupils who solved problem 14. The difference between the percentages is 2 in favor of the imaginative member of the pair. The rest of the table may be read accordingly. For each test the first column shows the number of unpaired solutions of the conventional problems; the second column shows the number of unpaired solutions of the imaginative problems; the third column shows the percentages of unpaired solutions to total solutions of conventional problems; the fourth column shows the corresponding percentages for the imaginative problems; and the fifth column shows the differences in percentages. In the lower rows the averages are shown. The first set of averages is of all the problems; the second is of the six pairs being given special attention.

TABLE 19

FREQUENCIES AND PERCENTAGES OF UNPAIRED SOLUTIONS

Problems	Problem Test I					Problem Test IV				
	C	I	% C	% I	C–I	C	I	% C	% I	C–I
AA 9–14	27	34	9	11	−2	37	134	9	28	−19
AS 15–10	20	47	8	16	−8	34	56	7	11	−4
AM 7–4	70	44	35	25	10	158	51	35	15	20
AD 1–6	17	66	12	34	−22	59	74	17	21	−4
SS 11–16	30	31	15	16	−1	110	52	26	14	12
SM 3–12	48	16	48	24	24	90	34	52	29	23
SD 19–2	43	17	22	10	12	99	35	24	10	14
MM 13–18	32	77	16	32	−16	43	167	16	42	−26
MD 17–8	35	26	17	13	4	73	43	19	12	7
DD 5–20	51	54	30	32	−2	65	98	22	26	−4
Average	373	412	19	21	−2	768	744	21	20	1
Average (6 pairs)....	215	231	18	19	−1	533	311	21	14	7

The possibility of rendering a conventional problem easier of solution through a lengthening of its statements is indicated by the data for the six pairs, namely, the AS, AM, AD, SS, SD, and MD pairs. The last row of the table shows for Problem Test I about the same number of unpaired solutions of the conventional problems as of the imaginative, and about the same percentages. Considerable differences in favor of the longer conventional problems are shown in the corresponding averages for the six selected pairs of Problem Test IV.

The Point Score

Excluding the four pairs of problems in question, point scores were computed for Problem Test IV in the manner described in the foregoing chapter used for determining the point scores for Problem Test I. It is to be kept in mind that the point scores represent the differences between the C (conventional) and I (imaginative) points when only six pairs of problems are taken into account. A C–I point score thus represents the superiority in points of the six longer conventional problems over the corresponding six imaginative problems.

Distribution of the Point Scores

The distribution of the point scores of 630 pupils in Problem Test IV is shown in Table 20. The table may be read: 5 pupils were more successful in the imaginative problems by 4 points, 10 pupils made a higher score by 3 points, 35 by 2 points, and so on. The figures to the right of the zero column indicate the numbers

TABLE 20

The Distribution of the Point Scores in Problem Test IV—6 Pairs

City	The Point Scores											Total	Average
	−4	−3	−2	−1	0	1	2	3	4	5	6		
Huntington ...	1	8	17	40	90	66	35	18	14	2	2	293	.56
Sheboygan	3	1	11	34	90	48	32	25	7	1	1	253	.61
Racine	1	1	7	10	30	19	4	7	3	1	1	84	.41
Total	5	10	35	84	210	133	71	50	24	4	4	630	.58

of pupils who were more successful in the conventional problems. The table may be read in another way. Exactly one-third of the scores recorded show no difference in pupil-responses to the two types of problems; 134, or 21 per cent, of the scores show a superiority in favor of the imaginative problems of one to four points; and 286, or 45 per cent, show a superiority in favor of the conventional problems of one to six points. As a whole, the table shows that the responses to the conventional problems were more favorable than were the responses to the imaginative problems. When compared with the distributions of the point scores in Problem Test I, which are given in Tables 6 and 7, the present distribution of point scores shows a clear gain which has been brought about by a lengthening of the conventional problems to include statements designed to make clearer the strictly mathematical elements.

In spite of the fact, however, that a clear gain is shown, the impressive fact remains that the large majority of the pupils are not helped one way or the other by an amount greater than one point. For two-thirds of the pupils concerned it may be said that the one type of problem statement is just as good as the other. Their practice in the recognition of the fundamental operations and thus their training toward a better acquaintance with the significance and use of these operations are not furthered by a modification of the usual conventional form of statement of the problems which provide the practice materials.

Explanation of the Distribution

On the other hand, one can not lose sight of the fact that an appreciable percentage of the pupils appears to have been helped by the one or the other type of problem to be found in Problem Test IV. The reasons why certain pupils appear to have been helped and others to have been hindered by the longer conventional type of problem should be sought. It is the experience of teachers that modifications of teaching methods and of practice exercises must frequently be made for the purpose of giving special help or a special form of training to the few who differ more or less from the average in ability and preparation, and that the extent and nature of such modifications are usually determined by the amount and kind of variations in given cases from normal expectations. It is quite as important to study the exceptional demands of the few as it is to provide for the usual demands of the many.

Differences in ability to read, in intelligence, in carefulness of work, in ability in reasoning in arithmetic, and in ability in the fundamentals of arithmetic were considered as possible explanations of differences in the point scores. By means of the Collateral Tests, which have been described in Chapter III, scores in reading, in intelligence, in carefulness, in reasoning, and in the fundamentals were obtained from the 293 pupils in the Huntington 6B grades.

Comparisons of these scores with their corresponding C–I point scores follow.

The Influence of Reading Ability

The influence of reading ability is shown in Table 21. The table shows a double distribution. From left to right the point scores are shown, and from bottom to top the scores in reading are shown. Except along the column for the zero point scores there is to be seen no marked tendency for the scores to cluster at any particular point. The pupils who are poor in reading appear on

TABLE 21

The Relation Between Reading Ability and the Point Scores in Problem Test IV

Scores in Reading	The Point Scores										
	−4	−3	−2	−1	0	1	2	3	4	5	Total
13.......				3	8	5	3		1		20
12.......		2	2	3	6	4	4	1	1		23
11.......			3	2	4	2		1	1		13
10.......			1	3	8	1	3		2		18
9.......			1	3	14	1		3	1		23
8.......				5	12	10	4	5	2		38
7.......		2	2	8	11	15	4	1	2	1	46
6.......			1	5	12	9	6	4		2	39
5.......		3	3	3	7	7	1	1	1		26
4.......		1	1	2	2	7	4	1			18
3.......	1		3	2	3	4	1	1	3	1	19
Total ...	1	8	17	39	87	65	30	18	14	4	283

$r = -.04$

the whole to be helped by the longer conventional problems. The same may be said of the pupils who are of average ability in reading, and of those who are good in reading. The conclusion to be drawn is that no relation exists between ability in reading and the point scores. The coefficient of correlation, —.04, computed by the Pearson product-moment formula, bears out the conclusion. To the extent that any given pupil is helped in attacking one type of problem by his ability to read, it may be said that he is helped correspondingly in attacking the other.

The Influence of Intelligence

The relation between the scores in intelligence and the point scores is shown in Table 22. The distribution of the scores in the table and the computed coefficient of correlation, —.09, show no relation between the pupil's score in an intelligence test and his point score in Problem Test IV. One may conclude that a given pupil will be found to be about equally intelligent in his attack upon the two types of problems represented. When a difference exists between his scores in the two types of problems, such difference must be sought in an examination of influencing factors other than the pupil's intelligence.

TABLE 22

The Relation Between Intelligence and the Point Scores in Problem Test IV

Intelligence Quotients	The Point Scores										
	−4	−3	−2	−1	0	1	2	3	4	5	Total
135–144					3						3
125–134					4	1	1				6
115–124		1	2	2	10	4	3	1			23
105–114		1	1	6	14	14	7	2	5		50
95–104		2	2	9	16	11	4	4	1		49
85–94		2	5	11	15	9	6	5		1	54
75–84		1	4	7	9	12	5	3	6	3	50
65–74		1	1	2	3	6	4	3	2		22
64 and Less	1				2	1		1			5
Total...	1	8	15	37	76	58	30	19	14	4	262

$r = -.09$

The Influence of Carefulness of Work

The influence of carefulness of work is shown in Table 23. The table shows that of the pupils who were inclined to be careless in their methods of work there was a higher percentage helped by the conventional problems than there was of those who were inclined to be careful. The figures seem to indicate that the careful pupil is on the whole about equally successful in his attack upon the two types of problems, while the careless pupil appears to find some assistance in the extra statements which are included in the conventional problems. The fact, however, that many of the careless

TABLE 23

THE RELATION BETWEEN CAREFULNESS OF WORK AND THE POINT SCORES IN PROBLEM TEST IV

SCORES IN CAREFULNESS OF WORK	THE POINT SCORES										
	−4	−3	−2	−1	0	1	2	3	4	5	TOTAL
90–99					2						2
80–89					9	1	1				11
70–79		1	3	3	12	8	1				28
60–69		1	2	8	11	7	7	6	1		43
50–59		2	4	9	14	12	9	4	3	1	58
40–49		3	5	8	10	11	5	1	3	1	47
30–39		1		6	8	12	5	5	3	1	41
20–29			1	3	8	6	2	3	2	1	26
10–19	1				2	1			2		6
Total...	1	8	15	37	76	58	30	19	14	4	262

$r = -.15$

pupils are not helped at all by the extra statements is indicative that the careless pupil takes no account of the extra help which they are intended to provide. The correlation, $r = -.15$, is so small as to be unreliable. It indicates no appreciable relationship.

THE INFLUENCE OF REASONING ABILITY

The influence of reasoning ability in arithmetic is shown in Table 24. Coefficients of correlation were computed both when the zero row of scores in the Stone Reasoning Test was included and when the zero row was not included. In the former case, the coefficient found was $r = -.03$; in the latter case, it was $r = -.06$. Both the table and the computed coefficients indicate that the apparent superiority of the one or the other type of problem in the case of any pupil cannot be explained by the pupil's ability to solve problems or by the lack of it.

TABLE 24

THE RELATION BETWEEN REASONING ABILITY IN ARITHMETIC AND THE POINT SCORES IN PROBLEM TEST IV

SCORES IN REASONING TEST	THE POINT SCORES										
	−4	−3	−2	−1	0	1	2	3	4	5	TOTAL
10 and over...					5						5
9........					1						1
8........					1		2				3
7........					2		1				3
6........		1			3	4		1			9
5........		1		3	8	3	2	1		1	19
4........		1	5	4	8	7	6	3		1	35
3........		1	2	10	19	12	9	2	2		57
2........			1	7	6	10	2	4	7	2	39
1........		2	4	6	10	8	1	4	3		38
0........	1	2	3	7	13	14	7	4	2		53
Total....	1	8	15	37	76	58	30	19	14	4	262

$r = -.03$, when the zero scores in reasoning are included.
$r = -.06$, when the zero scores in reasoning are not included.

TABLE 25

THE RELATION BETWEEN ABILITY IN THE FUNDAMENTALS AND THE POINT SCORES IN PROBLEM TEST IV

SCORES IN THE FUNDAMENTALS	THE POINT SCORES										
	−4	−3	−2	−1	0	1	2	3	4	5	TOTAL
225 and over				1	2	1	1				5
200–224				1	3		1		2		7
175–199		1		1	6	4	1	1	1	1	16
150–174			3	1	14	6	2	3	2		31
125–149		1	3	8	11	10	12	2	2		49
100–124		2	6	7	22	20	6	6	4	2	75
75–99	1	1	2	12	10	11	6	6	2	1	52
50–74		3	1	3	9	5	1	1	1		24
49 and less				1		1					2
Total....	1	8	15	35	77	58	30	19	14	4	261

$r = .09$

The Influence of Ability in the Fundamentals

The influence of ability in the fundamental operations of arithmetic is shown in Table 25. The table presents a hint that when the pupil is poor in the fundamentals and must give considerable attention to the mechanics of computation, he is just as likely to be helped by the one type of problem as by the other, but that the pupil who has acquired a mastery of the fundamental operations involved has a better chance to take advantage of any help a special statement of a problem may have to offer. The coefficient of correlation, $r = .09$, is too small and too unreliable, however, to give any particular weight to such a conclusion.

Summary of the Correlations

The correlations which have just been presented are summarized in Table 26 with similar correlations when the point scores were derived from the scores on the ten pairs of problems. The lack of influence of the various factors which has just been discussed is confirmed by the corresponding correlations which are obtained when the point scores from the ten pairs of problems are taken into account. The distribution of the point scores shown in Table 20 indicated only slight justification for the longer conventional statement of problems for all of the pupils tested. The present table shows the justification for no increase for any particular grouping of pupils.

TABLE 26

Coefficients of Correlation Between the Results of the Collateral Tests and the Point Scores

Collateral Tests	C–I Point Scores from 6 Pairs of Problems	C–I Point Scores from 10 Pairs of Problems
Burgess Reading	— .04	— .17
Otis Intelligence	— .09	— .11
Carefulness	— .15	— .05
Stone Reasoning (zero scores included)	— .03	....
Stone Reasoning (zero scores excluded)	— .06	....
Cleveland Fundamentals	.09	.02

Summary

1. The results of Problem Test IV show negligible differences between the totals of responses to the two types of problems.

2. The responses of the children in the cities of Sheboygan and Racine are on the whole slightly better than the responses of the Huntington children, but are withal so similar as to indicate that the responses of the Huntington children are not erratic, but fairly typical and representative.

3. Analysis of the responses to the various pairs of problems shows that a lengthening of the conventional manner of statement proved to be helpful in the case of six of the problems, and not helpful in the case of four of the problems. The danger of rendering the terse, explicit statement of a problem less easy to grasp through an overloading with words is thus indicated.

4. In the case of the six problems, the improvement in responses is large enough to indicate some degree of significance.

5. A study of the frequencies of unpaired solutions shows that the improvement in the case of the six problems is large enough to warrant further inquiry into the reasons for improvement.

6. The distribution of the point scores upon the six pairs of problems is a further indication of a clear gain in number of responses through a lengthening of the statements of the conventional problems. In spite of the clear gain, however, a large percentage of the pupils tested were neither helped nor hindered by the longer conventional statements.

7. The correlations of the point scores with scores in reading, in intelligence, in carefulness of work, in arithmetical reasoning, and in the fundamentals show that the degree of helpfulness of the longer conventional statements over the imaginative statements can not in any case be ascribed to differences in either of the activities or abilities named. The coefficients of correlation in every case are so low as to be considered negligible and unreliable.

In brief, it may be said that while the pupils of the intermediate grades may be expected to be helped in certain instances by the longer conventional type of statement of problems, the amount of such helpfulness in any case is largely a matter of chance.

CHAPTER VI

SIMILARITIES IN PUPIL-RESPONSES

The studies of the differences in the responses of pupils to the conventional and imaginative problems in the two chapters immediately preceding revealed no significant differences, and no reasons for the varying amounts of the differences found other than the operation of chance. The present chapter will reverse the point of view which has heretofore been maintained in the study of the two types of problems by considering the similarities in pupil-responses to the two contrasting types of problems under investigation.

The Frequency of Paired Solutions

The method of studying the similarities in pupil-responses which first suggests itself is the method of counting in the case of each pair of problems the occurrences of paired solutions and comparing the number found with the total number of solutions. A shorter method of arriving at the same results is to count in the case of each pair of problems the number of unpaired solutions, that is, the times either problem was solved when the other was missed. Deducting this number from the total for the pair, which is given in the case of Problem Test I in Table 2 and in the case of Problem Test IV in Table 16, gives the number of paired solutions. Dividing the number of paired solutions by the total gives the percentage of paired solutions.

Percentages of Paired Solutions

The percentages of paired solutions for the various pairs of problems in Problem Test I are given in Table 27. The table shows that in the 5B grades 83 per cent of the total solutions of the AA pair are paired solutions, in the 5A grades 95 per cent, in the 6B grades 95 per cent, and that the average percentage is 90. The average in each instance was found by dividing the total solutions which are paired by the total paired and unpaired.

Similar percentages for the problems of Problem Test IV are

TABLE 27

PERCENTAGES OF PAIRED SOLUTIONS IN PROBLEM TEST I

PROBLEMS	GRADES			AVERAGE
	5 B	5 A	6 B	
AA 9–14	83	95	95	90
AS 15–10	84	94	89	88
AM 7–4	64	75	71	69
AD 1–6	63	83	80	76
SS 11–16	73	92	89	84
SM 3–12	59	66	61	61
SD 19–2	75	87	89	84
MM 13–18	66	84	77	75
MD 17–8	76	92	87	85
DD 5–20	70	66	68	68
Average	74	85	83	80

shown in Table 28. Table 28 shows the percentages by cities, while Table 27 shows the percentages by grades. There is a hint in Table 27 that the better the responses to the two types of problems being studied, the higher are the percentages of paired solutions. (It is to be remembered that the 5A pupils who are represented made higher scores in Problem Test I than did the 6B pupils.) No such hint, however, is to be found in Table 28.

TABLE 28

PERCENTAGES OF PAIRED SOLUTIONS IN PROBLEM TEST IV

PROBLEMS	HUNTINGTON	OTHER CITIES	AVERAGE
AA 9–14	81	79	80
AS 15–10	91	91	91
AM 7–4	71	75	74
AD 1–6	74	86	81
SS 11–16	81	79	80
SM 3–12	59	55	57
SD 19–2	80	83	82
MM 13–18	69	68	69
MD 17–8	83	85	84
DD 5–20	78	73	75
Average	79	79	79

The impressive fact presented in the two tables is the high percentage of similarity in the responses of the pupils tested to the two types of problems. In every case it is to be noted that the pairs of problems evoked from the pupils responses which were more nearly alike than they were different.

THE INDEX OF SIMILARITY

Another method of studying the similarity in the responses of pupils to the two sets of problems is to compute for each pupil a similarity score, or a percentage of similarity, which, while not absolute, may be considered an index of the similarity of his responses. Thus in determining an individual score for each pupil, one is provided with the means of studying some of the factors which contribute to, or accompany, a similarity of response.

The method of computing the index of similarity is as follows: (1) Count the number of paired, or similar, solutions. (2) Count the number of *different* solutions, that is to say, the number of pairs of problems one or both members of which have been solved. (3) Divide (1) by (2) to find the percentage. (4) Count the number of paired points. (5) Count the number of pairs of points, one or both of which have been gained. (6) Divide (4) by (5) to find the percentage. (7) Average (3) and (6) to find the similarity score, or the index of similarity.

Let us suppose that a pupil's paper has been graded and marked thus:

AA	AS	AM	D	SS	X	SD	X	M	D
9	15	7	1	11	3	19	13	17	5
14	10	4	6	16	12	2	18	8	20
AA	AS	A	A	X	X	X	MM	D	D

The numbers shown are the numbers of the problems. The lettering above the odd numbers indicates the points gained in the conventional problems; the lettering below the even numbers indicates the points gained in the imaginative problems. Two letters together show that the particular problem to which they belong has been solved.

The method of computing the similarity may be illustrated thus: (1) Both members of the AA and AS pairs have been solved.

There are thus 2 paired solutions. (2) The pupil performed 6 *different* solutions of the various pairs, namely, AA, AS, AM, SS, SD, and MM, that is, he recognized either once or twice what was demanded by 6 different types of problems. (3) Since 2 divided by 6 is 33 per cent, the recognitions of the pairs are 33 per cent of the total different recognitions. (4) There are 6 pairs of points shown in the markings, 2 pairs for the AA problems, 2 for the AS, 1 for the AM, and 1 for the DD problems. (5) There are 17 different recognitions shown, as follows: 2, A and A, in the 9-14 pair; 2, A and S, in the 15-10 pair; 2, A and M, in the 7-4 pair; 2, A and D, in the 1-6 pair; 2, S and S, in the 11-16 pair; none in the next; and 2 each in the 19-2, 13-18, and 17-8 pairs, and 1, D, in the 5-20 pair. (6) Six divided by 17 is 35 per cent, which is the similarity score for the points. (7) Averaging 33 and 35 gives 34, which is set down as the index of similarity of the paper illustrated.

The question which the index of similarity answers is: Of the total number of different recognitions made by a given pupil, what per cent are similar recognitions? The average of the percentages serves to provide a score for the poorer pupils who were able to solve only a few, or none, of the problems, while for the better pupils the scores for the problems are affected but little, if at all, by the scores for the points.

The similarity scores for the pupils who took Problem Test IV were computed in the manner indicated above, that is, without excluding the four pairs of problems. Table 28 indicates that the percentages of paired solutions of the AA, SM, MM, and DD pairs are not very much lower than the percentages of paired solutions of the other pairs.

Distributions of the Similarity Scores

The distributions of the index of similarity scores in Problem Test I and Problem Test IV are shown in Tables 29 and 30. To aid in studying the distributions, they have been correlated with their respective distributions of the numbers of problems solved in the two tests. The tables show a high degree of relationship between the number of problems solved and the similarity of the solutions. The coefficients of correlation are $r = .84 \pm .009$ for the

TABLE 29

THE RELATION BETWEEN THE INDEX OF SIMILARITY SCORES AND THE NUMBER OF PROBLEMS SOLVED IN PROBLEM TEST I

PROBLEMS SOLVED	INDEX OF SIMILARITY SCORES											TOTAL
	0–9	10–19	20–29	30–39	40–49	50–59	60–69	70–79	80–89	90–99	100	
19–20										22	17	39
17–18								12	22	7	4	45
15–16						1	6	26	11	3	2	49
13–14						4	10	9	7	1	2	33
11–12				1	6	7	11	8		1		34
9–10				4	5	8	4	7	1			29
7–8			1	4	10	5	5	5		1		31
5–6			3	8	3	8	4	1			1	28
3–4	4	6	2	7	7	6		1				33
1–2	13	19	9	1	1		2	4	2	1	2	54
0	13	8	12	5		20						58
Total....	30	33	27	30	32	59	42	73	43	36	28	433

$r = .84 \pm .009$

TABLE 30

THE RELATION BETWEEN THE INDEX OF SIMILARITY SCORES AND THE NUMBER OF PROBLEMS SOLVED IN PROBLEM TEST IV

PROBLEMS SOLVED	INDEX OF SIMILARITY SCORES											TOTAL
	0–9	10–19	20–29	30–39	40–49	50–59	60–69	70–79	80–89	90–99	100	
19–20										25	29	54
17–18								26	46	17	9	98
15–16						1	12	39	44	4	2	102
13–14					2	23	17	22	16	1		81
11–12				2	10	16	17	6	4			55
9–10			3	6	13	5	7	10		1		45
7–8		1	2	11	12	6	7	3	1			43
5–6		3	4	7	7	8	8	2				39
3–4	1	13	7	6	6	6	1	3	3			46
1–2	7	17	5	3					1			33
0	14	5	5	3		7						34
Total....	22	39	26	38	50	72	69	111	115	48	40	630

$r = .85 \pm .008$

data in Table 29 and $r = .85 \pm .008$ for the data in Table 30. For those portions of the data of Problem Test IV which belong to the 293 pupils of Huntington, the coefficient of correlation is $r = .86 \pm .01$.

The high correlations which are indicated by the coefficients seem to show that the degree of similarity in a given case is due almost entirely to the number of problems which have been solved. Indeed, this is true for the pupils who were able to solve twelve or more of the twenty problems in a test. In such cases the similarity scores were bound to be high, due to the fact that there were only ten pairs of problems to solve. In other words, the greater the number of problems solved in the tests which were limited to ten pairs, the less became the possibility of dissimilarity in the responses. In a sense, therefore, the index of similarity score is not a measure of similarity of possible responses for the reason that the extent of the tests limits the possibility of performance for the better pupils.

On the other hand, it is to be kept in mind that since practice effects upon a given type of problems are limited in accordance with the law of diminishing returns, it is inadvisable to extend the practice exercises for the better pupils beyond reasonable limits. It is the procedure of the school, therefore, not to extend the practice exercises of a given type or degree of difficulty to the bounds of the abilities of the better pupils, but to offer such amounts as are reasonably limited in extent. In the school, then, the better pupils would be given limited lists of problems for their practice, and the similarity scores in exercises similar to the problems of the tests would be expected to be correspondingly high.

When one examines the scores of the poorer pupils in Tables 29 and 30, he can note that low degree of similarity does not always accompany a low problem score. The fact that some of these pupils made high similarity scores would indicate that the index of similarity is at least in part a true measure of similarity of response.

For example, Table 29 shows 20 of the 58 pupils who failed to solve any of the problems of Problem Test I making a similarity score of 50, and one-fifth of the pupils who solved one and two problems making similarity scores from 60 to 100. A high degree of similarity in the point scores, even when the number of problems solved was very low, is thus indicated for certain of the pupils. The coefficient of correlation for the lower two-thirds of Table 29, that is, for the problem scores ranging from 0 to 12, is $r = .51$. This considerably lowered coefficient of correlation indicates further that the index of similarity serves to measure the degree of similarity of responses to the two types of problems by establishing the fact that the similarity scores are not due entirely to the numbers of problems solved. The fact that the coefficient is so high, when considered in connection with the correlation when the scores of all the pupils are included, points to the conclusion that when a pupil has mastered a given type of problem situation, he can be expected to solve the problem whatever the manner of its statement.

The Influence of Reading Ability

The relation between reading ability and similarity of responses to the two types of problems in Problem Test IV is shown in Table 31. The table shows the related scores of 281 Huntington pupils.

The coefficient of correlation computed from the data in Table

TABLE 31

The Relation Between Reading Ability and Similarity in Responses to Problems of the Two Types

Scores in Reading	Index of Similarity Scores											Total
	0–9	10–19	20–29	30–39	40–49	50–59	60–69	70–79	80–89	90–99	100	
13					1	1	3	4	6	3	2	20
12			1	2		1	6	6	6		1	23
11	1	2		2			2	3	1	1	1	13
10		1		2	1		1	3	1	5	4	18
9	1			1	4	5	1	4	3	3	1	23
8	2	2	3	4	2	4	5	5	6	2	3	38
7	3	8	2		2	6	6	9	7	1	1	45
6	4	5	2	3	3	5	6	5	6			39
5	1	2	3	3	3	1	3	5	1	4		26
4	2	2	4	1	1	1	1	1	4		1	18
3	1	3	2	2	2	3	2	1	2			18
Total.	15	25	17	20	19	27	36	46	43	19	14	281

$r = .32 \pm .038$

31 is $r = .32 \pm .038$. The coefficient and the arrangement of the scores in the table indicate some degree of relationship, but not a marked dependence of similarity scores upon the varying abilities to read material of the kind represented by the paragraphs of the Burgess Reading Scale. To the extent that the child's ability in ordinary reading is carried over to the reading of his problems in arithmetic a relation between the two scores may be expected. On the other hand, the table would seem to indicate that a child may be able to attack two types of problems equally well, or equally poorly, as the case may be, and at the same time be either good or poor in ordinary reading activities. The coefficient obtained is about what may be expected when it is considered that factors like training in problem-solving and familiarity with problem situations may have more to do with a pupil's reaction to a problem than the degree of his ability to get meaning from printed material of the ordinary, story type.

The Influence of Intelligence

The influence of intelligence upon the similarity scores is shown in Table 32. The coefficient of correlation computed from this table is $r = .45 \pm .033$. The table indicates that the pupil with a high intelligence rating may usually be expected to attack one type of problem statement about as well as another, while the pupil with a lower rating may be either equally unintelligent in his attacks upon the problems of the two types, which would result in chance, or dissimilar, scores, or equally intelligent in a minor degree in his attacks, which would result in highly similar scores. The fact that some pupils with high intelligence ratings produced low similarity scores may be explained by the fact that answering the questions in an intelligence test and attacking problems in arithmetic are not for many pupils highly similar activities. For example, some pupils with good ability will attack an intelligence test with enthusiasm and look upon a test in solving arithmetic problems as an irksome undertaking. Such pupils will produce high scores in the one test and chance scores in the other. The fact that the percentage of the pupils with high intelligence ratings who made high similarity scores is larger than the percentage who made low scores indicates that the intelligent pupil may be

TABLE 32

THE RELATION BETWEEN INTELLIGENCE AND SIMILARITY IN RESPONSES TO PROBLEMS OF THE TWO TYPES

INTELLIGENCE QUOTIENTS	INDEX OF SIMILARITY SCORES											TOTAL
	0–9	10–19	20–29	30–39	40–49	50–59	60–69	70–79	80–89	90–99	100	
135–144											3	3
125–134							1	1	1	1	2	6
115–124				2	1	1	4	5	6		4	23
105–114	1	2		3	2	5	10	12	5	7	3	50
95–104		2	2	3	6	5	6	11	9	5	1	50
85–94	6	6	3	5	6	5	7	11	3	1		53
75–84	2	9	4	5	3	8	6	4	6	1		48
65–74	3	3	3	2	2	1	4	1	2	2		23
64 and less			3			1			1			5
Total..	12	22	15	20	20	26	38	45	33	17	13	261

$r = .45 \pm .033$

TABLE 33

THE RELATION BETWEEN CAREFULNESS OF WORK AND SIMILARITY IN RESPONSES TO PROBLEMS OF THE TWO TYPES

SCORES IN CAREFULNESS OF WORK	INDEX OF SIMILARITY SCORES											TOTAL
	0–9	10–19	20–29	30–39	40–49	50–59	60–69	70–79	80–89	90–99	100	
90–99										1	1	2
80–89		1						1	2	2	5	11
70–79				2		1		7	8	3	6	27
60–69		1	2	3	5	2	9	12	6	3		43
50–59	1	1	3	2	3	8	17	12	9	4		60
40–49	3	4	3	5	6	4	7	8	3	3	1	47
30–39	5	8	5	3	4	4	2	5	4			40
20–29	5	5	2	4	2	2	3		1	1		25
10–19			1	1		3		1				6
Total..	14	20	16	20	20	24	38	46	33	17	13	261

$r = .54 \pm .031$

expected to react with about equal success to dissimilar statements of similar type problem situations.

The Influence of Carefulness of Work

The influence of carefulness of work upon the similarity scores is shown in Table 33. The coefficient of correlation computed from this table is $r = .54 \pm .03$. The data of the table point to the conclusion that the pupils who are careful in their work are about equally careful in their attacks upon the two types of problems, while the pupils who are careless tend to produce chance scores in their attacks upon the problems. Only a small percentage of the pupils who are careless may be said to be similarly careless in their attacks upon the problems to the extent of producing high similarity scores. Careful work tends to a high degree of perfection in whatever one undertakes, and careless work tends to imperfections and sporadic successes. The comparative scores of the careless pupils are more subject to chance than they are to similar successes.

The Influence of Reasoning Ability

The influence of reasoning ability in arithmetic upon the similarity scores is shown in Table 34, which produces a coefficient of correlation of $r = .67 \pm .024$. The table shows only a few of the better pupils in reasoning ability making low similarity scores, and relatively few of the poorer ones in reasoning ability making high similarity scores. The table shows a marked degree of correspondence between low scores in reasoning and low scores in similarity, and between high scores in reasoning and high scores in similarity. In other words, it may be said that the pupil who is good in reasoning is about equally good in the two types of problems, while the pupil who is poor in reasoning is able to attack his problems with little success and to secure only chance successes. As in the correlation tables immediately preceding, it is shown also in Table 34 that certain pupils who are poor are similarly unsuccessful in their methods of attack upon the two types of problems and are thus able to secure high similarity scores.

TABLE 34

THE RELATION BETWEEN ABILITY IN ARITHMETICAL REASONING AND SIMILARITY IN RESPONSES TO PROBLEMS OF THE TWO TYPES

Scores in Reasoning	Index of Similarity Scores											Total
	0–9	10–19	20–29	30–39	40–49	50–59	60–69	70–79	80–89	90–99	100	
10.......											5	5
9.......											1	1
8.......								1		2		3
7.......								1		1	1	3
6.......						2	1	2	3		1	9
5.......			1		1	1	4	4	5	2	1	19
4.......			1	2	2	2	5	9	10	3	1	35
3.......		2	2	1		6	11	16	10	5	3	56
2.......		2	3	6	5	6	6	6	4	3		41
1.......	3	6		4	6	7	5	5	1			37
0.......	11	13	7	6	6	2	5	1		1		52
Total..	14	23	14	19	20	26	37	45	33	17	13	261

$r = .67 \pm .024$, when the zero scores in reasoning are included.
$r = .54 \pm .034$, when the zero scores in reasoning are not included.

TABLE 35

THE RELATION BETWEEN ABILITY IN THE FUNDAMENTALS AND SIMILARITY IN RESPONSES TO PROBLEMS OF THE TWO TYPES

Scores in the Fundamentals	Index of Similarity Scores											Total
	0–9	10–19	20–29	30–39	40–49	50–59	60–69	70–79	80–89	90–99	100	
225 and over							1		2	2		5
200–224					1	1	2	1			2	7
175–199		2		2		2	5	1	2		2	16
150–174				2	2	3	3	7	9	2	3	31
125–149	2	3	3	3	3	4	7	8	9	5	2	49
100–124	4	7	5	4	5	6	14	15	7	5	4	76
75–99	2	8	6	7	6	6	4	9	4	2		54
50–74	4	2	1	2	3	4		4		1		21
49 and less		1	1									2
Total..	12	23	16	20	20	26	36	45	33	17	13	261

$r = .34 \pm .038$

The Influence of Ability in the Fundamentals

The influence of ability in the fundamental operations in arithmetic upon the similarity scores is shown in Table 35, which produces a coefficient of correlation of $r = .34 \pm .038$. Table 35 shows the measures which are represented scattered fairly generously throughout its various parts. For the most part the pupils who are good in the fundamentals have succeeded in making high similarity scores. On the other hand, about as many of the poorer ones in the fundamentals made high similarity scores as made low ones. The former observation may be explained by the fact that success in problem-solving is usually accompanied by a reasonable degree of mastery of the fundamentals; the latter observation may be explained by the fact that there are to be found in the schools pupils who are poor in the fundamentals because of a lack of training but who are at the same time careful in their work and keen in their attack upon problems which come within their grasp.

Summary of the Correlations

The correlations which have just been presented are summarized in Table 36. The coefficients indicate low relationship of the similarity scores with reading ability and ability in the fundamentals, moderate relationship with intelligence, and marked relationship with carefulness of work and reasoning ability in arithmetic.

TABLE 36

Coefficients of Correlation Between the Results of the Collateral Tests and the Index of Similarity Scores

The Collateral Tests	Values of r
Burgess Reading	.32 ± .038
Otis Intelligence	.45 ± .033
Carefulness	.54 ± .03
Stone Reasoning	.67 ± .024
Cleveland Fundamentals	.34 ± .038

Similar indications of relationship may be shown in another way. The limited extent of the problem tests, which has been referred to in an earlier section of this chapter, itself produces high similarity scores for the better pupils, whether or not the pupils otherwise would be highly similar in their responses. Likewise, it may be shown that the better pupils in the collateral tests would be expected to make high similarity scores, not primarily because they have higher rankings in these tests, but because high ratings in these tests go with high ratings in the responses to the problem tests. This latter fact is indicated in the coefficients of correlation between the collateral test scores and the problem scores in Problem Test IV, which are shown in Table 37. For the reasons just given, the comparisons for the poorer pupils, who for the most part made the lower scores in the problem test, are the significant comparisons.

TABLE 37

COEFFICIENTS OF CORRELATION BETWEEN THE RESULTS OF THE COLLATERAL TESTS AND THE PROBLEM SCORES IN PROBLEM TEST IV

THE COLLATERAL TESTS	VALUES OF r
Burgess Reading	.32 ± .043
Otis Intelligence	.42 ± .04
Carefulness	.53 ± .036
Stone Reasoning	.70 ± .024
Cleveland Fundamentals	.38 ± .04

Such comparisons may be facilitated through a regrouping of the data which have been presented in Tables 31 to 35.

Table 38 shows the data of Tables 31 to 35 regrouped for each table into four divisions or quadrants. In making the arrangements each of the two distributions shown in each table was divided at or near its mean. These approximate mid-points were: At the score of 60 for the similarity scores; between 7 and 8 for the scores in reading; at 95 for the intelligence scores; at 50 for the scores in carefulness of work; between 2 and 3 for the scores in reasoning; and at 125 for the scores in the fundamentals.

Table 38 shows the arrangements into quadrants. It may be read thus: (1) Of the better half of the pupils in reading the relation of poor to good in similarity scores was as 43 to 92; of the poorer half in reading the relation was as 80 to 66; (2) of the better half of the pupils in intelligence the relation of poor to good in similarity scores was as 35 to 97; and so on.

TABLE 38

QUADRANT DISTRIBUTIONS OF THE COLLATERAL TEST SCORES AND THE INDEX OF SIMILARITY SCORES

60

8/7	43	92	8/7
	80	66	

60

(1) Reading

60

95	35	97	95
	80	49	

60

(2) Intelligence

60

50	35	108	50
	79	39	

60

(3) Carefulness

60

3/2	23	108	3/2
	93	37	

60

(4) Reasoning

60

125	33	75	125
	84	69	

60

(5) Fundamentals

Examining the lower halves of the quadrant distributions, which show the lower halves of the scores in the collateral tests, it may be seen that ratio of high scores in similarity to low scores approaches unity in parts (1) and (5), and that the ratios are lower in parts (2), (3), and 4), in the order given. Stating these ratios in percentages, it may be said that when in each case the poorer half of the pupils in the collateral tests are considered, they are found to be divided into good and poor in index of similarity scores as follows: In reading, the number who are high in similarity of response is 83 per cent of the number who are low; in intelligence, the corresponding percentage is 61; in carefulness, it is 49; in reasoning, it is 40; and in ability in the fundamentals, the percentage is 82. Poor reading ability and poor ability in the fundamentals are thus found to affect the scores in similarity only slightly (almost as many of such poorer pupils make high similarity scores as make low ones), whereas low intelligence seems to influence toward chance responses—and, as a consequence, low similarity scores—and lack of carefulness of work and poor reasoning ability exert a marked influence toward chance responses. The careless pupil and the one who has developed little reasoning ability, such as is indicated by the Stone test, are hit-and-miss propositions in so far as their attacks upon the two types of problems are concerned; the ones with low scores in intelligence may frequently be found to attack the two types of problems with about equal degrees of skill; and the pupils who read poorly or who are poor in the fundamentals are divided about equally into those who are haphazard in their attacks upon the two types of problems and those who are similarly successful, or unsuccessful, in their attacks.

Summary

1. The percentages of paired solutions of the members of the various pairs of problems of Problem Tests I and IV show that the pupils who were tested were more alike in their responses to the problems of the two types than they were different.

2. The distributions of the index of similarity scores lend further emphasis to the fact that responses were more similar than different.

3. The index of similarity scores correlates high with the total of problems solved. A more impressive fact, however, is the large number of high similarity scores which were made by pupils who solved but few problems.

4. The correlations of the index of similarity scores with the scores in the Collateral Tests produce in the case of reading and the fundamentals coefficients which are not high enough to be significant, and, in the case of intelligence, carefulness of work, and arithmetical reasoning coefficients which show a moderate degree of relationship. The pupils who were good in the Collateral Tests as a rule made high similarity scores; the pupils who were poor made both low and high similarity scores. Poor reading and low ability in the fundamentals affected the similarity scores the least; low intellectual keenness, carelessness of work, and poor arithmetical reasoning operated to produce chance successes in the problem tests.

In brief, there is clear indication in the results which have been presented in the present chapter that the pupil who has mastered a certain type of problem situation is able to deal with such a situation whatever the manner of its presentation. Unless the situation has been hidden in a mass of useless verbiage, the pupil can recognize the situation he knows, whatever its dress or its association.

CHAPTER VII

THE INFLUENCE OF TIME AS A FACTOR

The findings which have been presented in the three chapters immediately preceding have led to the conclusions (1) that the problem statement of the imaginative type possesses no element of superiority over the problem statement of the conventional type, and (2) that the problem statement of the so-called "longer" conventional type is slightly, but not significantly, superior to the problem statement of the imaginative type. The findings were those which were secured from problem tests which had been given under conditions in which the time element was not a factor other than that of providing opportunity to all the pupils tested to attack all of both types of problems which were under comparison. The present chapter will undertake a comparison of conventional and imaginative problems under conditions in which the time element is a constant factor. The question to be answered by this chapter is: What are the differences in the responses of pupils to the two types of problems when the time allowance for attacking each type is constant?

Problem Test V

The nature and method of administration of Problem Test V have been presented in Chapter III. The problems in the two parts, Test A and Test B, are assumed to be equal in difficulty. They all deal with situations with which pupils in the fifth and sixth grades are familiar, and they each require a common two-step operation, namely, subtraction and multiplication or a double multiplication and subtraction. Since the pupil's score in either test was determined by the choice of operation, rather than the accuracy of operation, the operations required may be assumed to impose demands equal in difficulty. Moreover, the time element is constant. There remains, therefore, the one variable factor, namely, the manner of statement of the problems.

The Influence of Training Within the Tests

While the factors which will influence the results secured, such as kind of problem, kind of situation, and operations involved, are theoretically constant, there develops out of the test itself an influence which may be considered either as a variable factor, or as one which renders variable the theoretically constant factors. This is the influence of the training which may be carried from one part of the test to the other. Because of the similarity of the problems, it is to be expected that with the solving of each problem the pupil will be better prepared to solve the next one, and that the pupil will carry over from the part of the test which is taken first some degree of training which will be helpful in the part which is taken last. The results which were secured confirm this expectation, and justify the method used to equalize the differences which may be due to the training within the test.

The method of equalizing the differences due to training within the test was (1) that of varying the order of giving Part A and Part B of the test, giving Part A first followed by Part B to one group of pupils and Part B first followed by Part A to a corresponding group; and (2) that of equalizing the scores which were obtained. The method may be illustrated in the presentation and treatment of the results.

The Results of Problem Test V

During the early weeks of April, 1928, Problem Test V was given to 264 pupils of the 6B grades in the Huntington, West Virginia, Public Schools. The pupils tested were the ones who had already taken Problem Test IV and the Collateral Tests.

Table 39 shows the results of Problem Test V for the schools in which the order of the parts was Test A first followed by Test B. This order is indicated by the legend, A1 B2. Table 40 shows the results for the schools in which the order was Test B first followed by Test A. This order is indicated by the legend, B1 A2.

Table 39 shows that when the test was given in the A1 B2 order the average number of problems solved in the A or conventional group is 2.74; in the B or imaginative group, the average is 2.14 problems. Table 40 shows that when the order of giving

TABLE 39

RESPONSES OT PROBLEM TEST V WHEN PRESENTED IN A1 B2 ORDER

SCHOOL	NUMBER OF PUPILS	PROBLEMS SOLVED	
		TEST A	TEST B
Marshall	21	78	63
Holderby	26	75	58
Enslow	26	59	48
Johnson	16	47	33
Miller 1	35	83	62
Guyandotte	29	77	64
Total	153	419	328
Average		2.74	2.14

the test was reversed the corresponding averages are 3.3 and 1.7. That pupils in the two groups are not widely disparate in ability is shown by the fact that the average of problems solved by A1 B2 group totals 4.9 and by B1 A2 group the average totals 5.0.

TABLE 40

RESPONSES TO PROBLEM TEST V WHEN PRESENTED IN B1 A2 ORDER

SCHOOL	NUMBER OF PUPILS	PROBLEMS SOLVED	
		TEST A	TEST B
Simms 1	23	73	34
Simms 2	15	48	23
Oley	21	88	49
Miller 2	24	76	38
Emmons	28	79	46
Total	111	364	190
Average		3.3	1.7

THE SUPERIORITY OF THE SHORTER PROBLEMS

The data of Table 39 may be totalled to show that the 152 pupils represented solved 91 more of the shorter conventional problems than of the longer imaginative problems, which averages .6 problem per pupil; and the data in Table 40 may be similarly totalled to show that the 111 pupils therein represented solved 174 more of the shorter conventional problems than of the longer imaginative problems, which averages 1.6 problems per pupil. Both tables show, further, that the 264 pupils were able to solve in ten minutes 265 more of the short, tersely-stated problems, which is an average of 1 problem more per pupil.

There is nothing in the results found to indicate a denial of the findings presented in preceding chapters to the effect that the pupils may be expected to do equally well on the two types of problems under unequal time conditions. The fact that the pupils who took Test V solved fewer imaginative problems does not indicate a less degree of ability to handle the longer statements. The only fact of significance is that to deal with the longer problems takes longer time. The question answered is not, how many problems of the one kind or the other are the pupils able to solve, but how many can they solve in a limited time?

Time is a factor in the work of the school. Practice periods must be limited in extent. The teacher's problem is to provide the greatest amount and the most effective practice in the time at hand. Shall the teacher, then, choose for the practice exercises of the pupils short problems or long ones, tersely-stated or wordy problems, conventional or imaginative problems? The answer in Tables 39 and 40 is that the shorter problems provide the greater amount of practice in a given length of time.

The Degree of Superiority

The degree of superiority of the shorter conventional problems for the pupils as a whole has been approximated at 1 problem per pupil. The degree of superiority in individual cases is more difficult to determine. The tables show the effect of practice within the test from A to B in the one case and from B to A in the other. Certainly the gross difference in the scores of one of the 153 pupils in A1 B2 group would not be comparable to the gross difference in the score of one of the 111 pupils in the B1 A2 group. For the sake of making comparable the differences, the method suggests itself of making a deduction from the B score in the former case, and from the A score in the latter case. The fault to be found with this method is that in certain individual cases no problems in Test B were solved, and in certain other cases no problems in Test A were solved.

The method chosen was an approximation. The average A-B problem score for the A1 B2 group is .6, and for the B1 A2 group it is 1.6. Comparable individual scores for approximate comparisons with the scores in the Collateral Tests and in Problem Test IV may be secured by reducing each individual A-B problem score for the B1 A2 group 1 problem. It is true that the A-B problem scores of the B1 A2 group may not have been influenced equally by the changed order of the tests. On the other hand, they are sufficiently comparable among themselves to be set down in a double distribution with other test scores with which they are to be compared, and when so set down they are as a whole one point removed from the corresponding scores of the A1 B2 group. To bring them back through a general reduction of one point would not vitiate an approximate comparison with other test scores undertaken to determine the coefficient of correlation. The same coefficient would result in the case of 111 A-B scores for the B1 A2 group whether they are left as originally found or subjected to a general reduction. But when used with the remaining 153 A-B scores in a comparison with others a general reduction would be necessary in order to produce a coefficient which the more closely approximates the actual. Accordingly, the A-B problem scores of the B1 A2 group were each reduced by 1, and were used later with the remaining 153 scores to determine causes of individual variations. The coefficients of correlation between the total scores thus obtained and the scores in the Collateral and Problem IV Tests will be shown to be so low and so unreliable as indicators of relationship as to justify the approximations just indicated.

Distributions of the Problem Scores

Table 41 shows the distribution of the problem scores in numbers of pupils solving given numbers of problems for the A1 B2 group, and Table 42 shows the corresponding distribution for the B1 A2 group. It may be seen that 70 pupils—44 in the A1 B2 group and 26 in the B1 A2 group—failed to solve any of the problems in the two parts of the test. Such zero scores were excluded from the comparisons.

TABLE 41

FREQUENCIES OF SOLUTIONS WHEN PROBLEM TEST V WAS GIVEN IN A1 B2 ORDER

Test A		Test B	
Number Solved	Number of Pupils Solving	Number Solved	Number of Pupils Solving
10	1	10	..
9	3	9	1
8	2	8	1
7	8	7	2
6	10	6	3
5	14	5	13
4	17	4	22
3	22	3	22
2	14	2	21
1	18	1	18
0	44	0	50

TABLE 42

FREQUENCIES OF SOLUTIONS WHEN PROBLEM TEST V WAS GIVEN IN B1 A2 ORDER

Test A		Test B	
Number Solved	Number of Pupils Solving	Number Solved	Number of Pupils Solving
10	..	10	..
9	1	9	..
8	3	8	1
7	1	7	..
6	14	6	3
5	12	5	5
4	11	4	8
3	20	3	11
2	14	2	29
1	9	1	16
0	26	0	38

DISTRIBUTION OF THE A-B SCORES

The distribution of the A–B Scores, with the scores of the B1 A2 group equated by a general reduction of 1, is shown in Table 43 Equal standings in the two parts of Problem Test V are indicated by the zero scores. Negative scores indicate larger number of solutions of the problems of Test B; and positive scores indicate larger number of solutions of the problems of Test A. It will be seen that the larger percentage of the 195 scores cluster at or about the zero point. However, the fact is significant that there are so many more + 1 scores than — 1 scores; likewise it is to be noted that the greater number of plus scores indicates a decided advantage of the conventional problems over the imaginative problems. Of the 195 pupils represented in the table, 111 are shown to have secured from one to five problems more practice when working with the shorter problems.

Reasons for the Distribution

Table 43 suggests an interesting question: Are the variations in the A–B scores due to variations in the abilities of the pupils? Perhaps the better readers will be helped less by the shorter problems because of their superior ability to get through with the extra reading demanded by the longer ones. Perhaps they will be helped more by the shorter problems, because their superior ability enables them to read and to solve more of both kinds of problems in comparison with the slower pupils. Similar conjectures can be made respecting intelligence, carefulness, arithmetical reasoning, and ability in the fundamentals.

TABLE 43

Distribution of the A–B Scores

A–B Scores	—2	—1	0	1	2	3	4	5
Number	8	22	54	54	32	16	7	2

Tables 44 to 48 show the correlations of the A–B problem scores with scores in the collateral tests. Table 49 summarizes the correlations. Practically no relationship is discoverable between degree of helpfulness of the shorter problems, and the varying abilities which are measured by the collateral tests. It appears that the conventional problems of Part A offer better opportunity for practice to all pupils; whether they take equal advantage of the opportunity does not appear to be a question of varying ability in the characteristics which are measured by the collateral tests.

TABLE 44

THE INFLUENCE OF READING ABILITY UPON THE DIFFERENCE SCORES IN THE TIMED TEST

SCORES IN READING	PROBLEM-DIFFERENCE SCORES								
	−2	−1	0	1	2	3	4	5	TOTAL
13.............	1		4	5	4	2	1	1	18
12.............	2	3	5	6	5				21
11.............		3	2	2			2		9
10.............		1	1	3	7	3			15
9.............		2	4	5	4				15
8.............	1	2	6	6	4	3	2	1	25
7.............		3	13	8	2	3			29
6.............	1	3	5	8	3	2	1		23
5.............	1	4	8	6		1			20
4.............	1		4	2	1	1			9
3.............	1	1	2	3	2	1	1		11
Total.........	8	22	54	54	32	16	7	2	195

$r = .09$

TABLE 45

THE INFLUENCE OF INTELLIGENCE UPON THE DIFFERENCE SCORES IN THE TIMED TEST

INTELLIGENCE QUOTIENTS	PROBLEM-DIFFERENCE SCORES								
	−2	−1	0	1	2	3	4	5	TOTAL
135–144				2					2
125–134		2		1	2	1			6
115–124		2	5	4	7	2			20
105–114	2	5	14	11	6	4	2		44
95–104	2	3	9	12	8	4	3		41
85–94	2	4	13	6	4	1	1	1	32
75–84	1	2	12	9	5	2	1		32
65–74	1	1	2	3	1	2			10
64 and less		1			1				2
Total.........	8	20	55	48	34	16	7	1	189

$r = .06$

TABLE 46

THE INFLUENCE OF CAREFULNESS OF WORK UPON THE DIFFERENCE SCORES IN THE TIMED TEST

Scores in Carefulness of Work	Problem-Difference Scores								
	−2	−1	0	1	2	3	4	5	Total
90–99		1				1			2
80–89			1	4	2	1			8
70–79	1	2	4	5	7	2	3		24
60–69		4	12	11	5	3			35
50–59	2	5	17	10	10	2	2		48
40–49	3	3	8	8	3	3			28
30–39	1	4	9	4	4	3	1	1	27
20–29		1	4	5	2	1	1		14
10–19	1			1	1				3
Total.........	8	20	55	48	34	16	7	1	189

$r = .09$

TABLE 47

THE INFLUENCE OF ARITHMETICAL REASONING UPON THE DIFFERENCE SCORES IN THE TIMED TEST

Scores in Reasoning	Problem-Difference Scores								
	−2	−1	0	1	2	3	4	5	Total
9..............				1	2	1			4
8..............		1	1			1			3
7..............				2		1			3
6..............			1	4	2	1	1		9
5..............		3	5	5	2	2	1		18
4..............	2	2	13	7	8		2		34
3..............	2	5	13	11	12	4	1		48
2..............	2	5	8	12	5	3	2	1	38
1..............	2	3	5	3	2	2			17
0..............		1	9	3	1	1			15
Total	8	20	55	48	34	16	7	1	189

$r = .16 \pm .05$

TABLE 48

THE INFLUENCE OF ABILITY IN THE FUNDAMENTALS UPON THE DIFFERENCE SCORES IN THE TIMED TEST

Scores in the Fundamentals	Problem-Difference Scores								
	−2	−1	0	1	2	3	4	5	Total
225 and over			2		1	2			5
200–224	1		3		1				5
175–199		1	2	2	2	3	1		11
150–174	1	6	7	8	6		2		30
125–149	2	3	8	13	9	4			39
100–124	2	5	20	10	12	7	3		59
75–99	1	5	9	10	3		1	1	30
50–74	1		4	5					10
Total.........	8	20	55	48	34	16	7	1	189

$r = .09$

TABLE 49

COEFFICIENTS OF CORRELATION BETWEEN THE A–B PROBLEM SCORES AND THE RESULTS OF THE COLLATERAL TESTS

The Collateral Tests	Values of r
Burgess Reading	.09 ± .05
Otis Intelligence	.06 ± .05
Carefulness	.09 ± .05
Stone Reasoning	.16 ± .05
Cleveland Fundamentals	.09 ± .05

SUMMARY

1. The present chapter compares the responses of pupils to problems of the conventional and imaginative types when the time element is a constant factor.

2. The training carried over from one part to another of Problem Test V was equalized (1) by varying the order of the parts and (2) by equating the differences in the problem scores.

3. The responses to the two parts of Problem Test V show that the conventional problems are superior to the imaginative problems as exercises for practice. More practice can be secured in a given time from problems of the former type than from problems of the latter type.

4. The degree of superiority in a given case does not appear to be related to the pupil's standing in reading ability, in intelligence, in carefulness of work, in arithmetical reasoning, or in ability in the fundamentals.

5. The conventional problems appear to offer superior opportunity for practice to all pupils.

CHAPTER VIII

THE INFLUENCE OF CHANCE

The suggestion has been made in the foregoing chapters that the differences which were found in the point scores in Problem Tests I and IV and in the problem scores in Problem Test V were chance differences. The reason for such suggestion is that the differences were found not to be related to the standings of the pupils in those abilities and qualities which were measured by the Collateral Tests and which are a part of or contributory to the ability to respond to the problem tests. Moreover, it was indicated in Chapter VI that when a pupil was found to be able to master any certain problem situation of the one type of statement, he was usually found to be the master of the corresponding problem situation of the other type of statement. Since one type of problem statement appeared on the whole to be about as influential in calling forth correct responses as the other, such differences as were observable were held to be more apparent than real and of the type which one would expect to occur through the influence of chance alone. It is the purpose of the present chapter to assemble the data which appear to give weight to such suggestion and conclusion.

Characteristics of the Chance Distribution

Distributions of occurrences which are subject to the influence of chance resemble with striking regularity a symmetrical curve. In all such distributions the measures or records of occurrences are largely concentrated near the middle of the scale. Moreover, they shade off symmetrically in both directions from the middle high point. The errors of refined measurement, e.g., astronomy, surveying, chemical analysis, etc., are chance errors and are so distributed. The axioms derived from experience with such distributions are: (1) Zero errors are most common. (2) Small errors are more frequent than large ones. (3) Positive and negative errors are equally frequent. (4) Very large errors do not occur. The curve of such distributions is known as the normal probability curve. It represents normal expectation as to the distribution of a large number of cases. It represents, moreover, normal expectation as to the influence of chance when conditions are such as are described by the statement: "The chances are equal."

When "the chances are equal," their opposing influences are commonly neutralized; less commonly, influence in the one or the other direction is slightly superior; still less commonly, the influence one way or the other is markedly superior. In the long run, the influences of chance in opposing directions are neutralized.

Let us illustrate by reference to the distribution of the C-I point scores in Problem Test I if they were due entirely to the influence of chance. It has been shown that the conventional and imaginative statements in this test possess, when time is disregarded, about equal merit as practice exercises. We shall assume, moreover, that a sufficiently large and representative group has been measured by the test. How then would chance affect the distribution? The answer follows:

1. Zero C–I point scores would be most common.
2. Small differences would occur more frequently than large ones.
3. Positive and negative differences would be equally frequent.
4. Very large differences would not occur.

Substituting Normal for Actual Distributions

The actual distribution of the C–I point scores in Problem Test I has been described in Chapter IV. How closely it approximates the normal, or chance, distribution can be shown by substituting for it its normal distribution. Diagram II illustrates the substitution.

The irregular curve, the frequency polygon, shows the actual distribution of the C–I point scores in Problem Test I. The highest ordinate at zero shows 89 pupils with a zero point score. Other ordinates show 62 pupils with —1 as the point score, 57 pupils with 1 as the point score, and so on. By computation we have found that the mean of the distribution is —.07 and that the standard deviation is 2.2. Our problem is to find the heights of the ordinates and their locations on the base line if the distribution were a normal one.

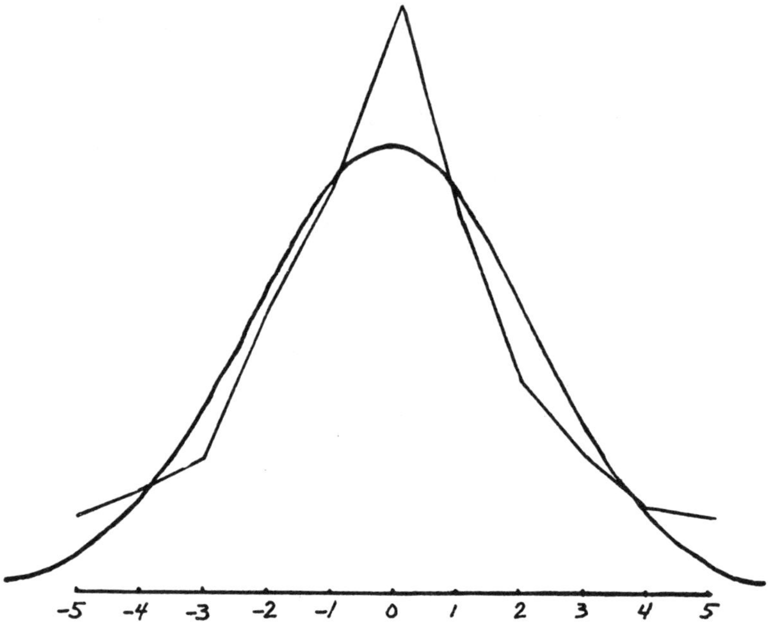

DIAGRAM II

ACTUAL DISTRIBUTION OF C–I POINT SCORES IN PROBLEM TEST I OF 375 FIFTH- AND SIXTH-GRADE PUPILS (PUPILS WITH ZERO SCORES NOT COUNTED) COMPARED WITH THE CORRESPONDING NORMAL DISTRIBUTION FOR THE SAME TOTAL NUMBER OF CASES AND THE SAME STANDARD DEVIATION

The computation is worked out by using the theorem that, in a normal distribution, the height of the maximum ordinate is equal to the number of cases divided by the standard deviation times the square root of 2π. This theorem expresses the relationship between the area of the normal surface of distribution and its maximum ordinate. Since the square root of 2π is just over 2.5, the theorem may be expressed thus:

$$y_o = \frac{N}{2.5\,\sigma}$$

The y_o, or maximum ordinate, was erected at the mean. The other ordinates were erected at points in multiples of $.25\sigma$ distance to the right and left, and were computed in terms of the middle ordinate and the standard deviation of the distribution, as follows:

$$y = y_o e^{\frac{-x^2}{2\sigma^2}}$$

The tops of the ordinates thus determined and located were connected by the smooth curve as shown in Diagram II.

Similar substitutions have been made for the C–I point scores of Problem Test IV and for the A–B problem scores of Problem Test V. These are shown in Diagrams III and IV, respectively.

Comparison of the Actual with the Normal Distributions

Diagrams II, III, and IV show how closely the actual distributions fit the normal distributions, or, in other words, how closely they approximate the distributions which may be due to chance alone. In Diagrams III and IV, the means are shown to be at the right of the zero difference scores. Foregoing chapters have presented the explanations of these divergences. In the former case, the explanation is that the longer conventional problems are slightly the superior of the imaginative problems; in the latter case, the explanation is that when time is constant the conventional problems are slightly the superior of the imaginative.

Closeness of fit with the normal distribution, when considered by itself, however, does not argue the operation of chance alone. A distribution of traits may approximate the normal, and yet at every point be shown as the resultant of causes which may not be designated as chance causes. For example, the heights of 10,000 men may be normally distributed. This fact does not prove the exclusive operation of chance. It may be shown for each height represented on the scale that the causes include inheritance, environmental conditions, food, health, etc. To be sure, it could be said that any one of the causes named, when operating throughout an extensive distribution, exerts a chance influence or operates similarly to chance. But height is not described as due to chance, but to definite influences.

Likewise, it may be argued that the point-and-problem-difference distributions, shown by the frequency polygons in Diagrams II, III, and IV, are not due to chance alone simply because they approximate the normal distribution. For example, let us suppose

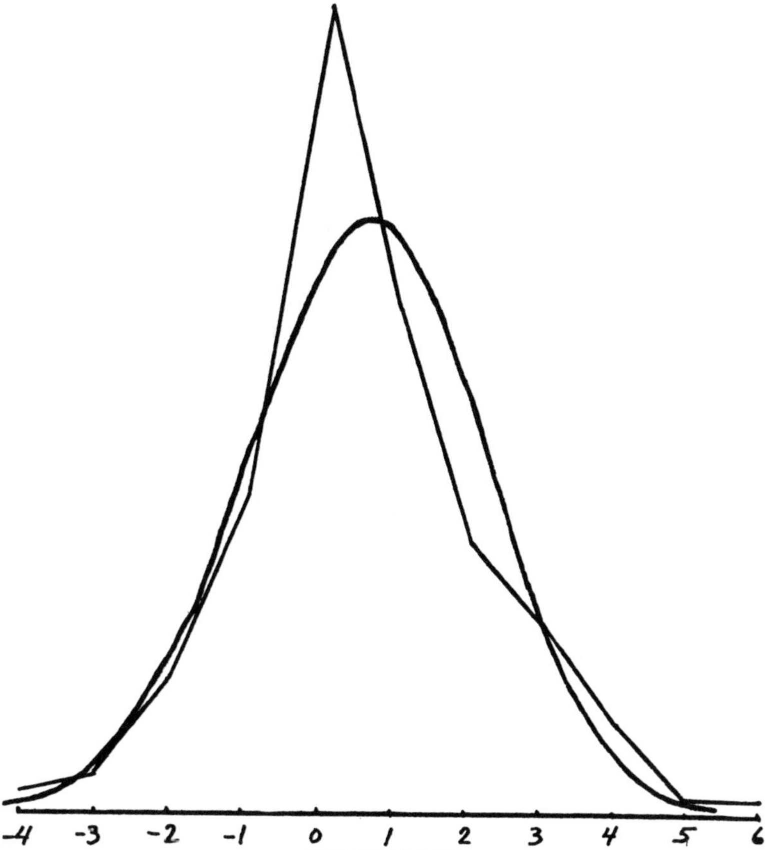

DIAGRAM III

ACTUAL DISTRIBUTION OF C–I POINT SCORES IN PROBLEM TEST IV OF 630 PUPILS COMPARED WITH THE CORRESPONDING NORMAL DISTRIBUTION FOR THE SAME TOTAL NUMBER OF CASES AND THE SAME STANDARD DEVIATION

that the point scores are due in a marked degree to reading ability —in other words, that reading ability affected scores in one type of problems more than in the other. In such a case, the difference scores, if enough were secured to be representative, would take the form of the normal curve, and would be set down as due, not to chance, but to reading ability. Moreover, in such a case, if the scores were few and not representative, their distribution would not approximate the normal curve. In the latter case, the distribution of the difference scores would depart from normality to

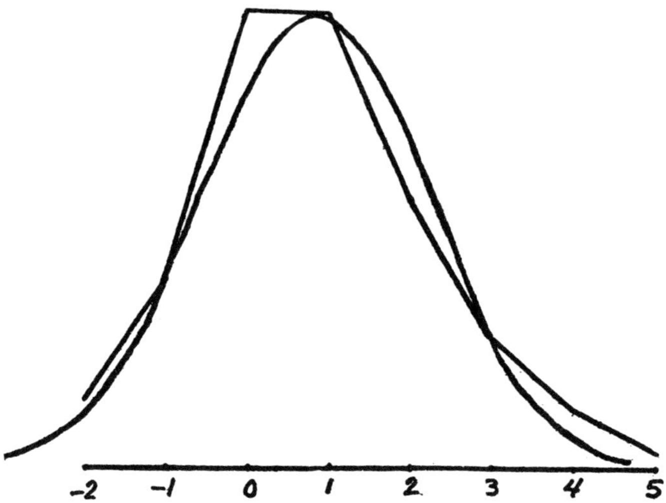

DIAGRAM IV

ACTUAL DISTRIBUTION OF A–B PROBLEM SCORES IN PROBLEM TEST V OF 195 PUPILS COMPARED WITH THE CORRESPONDING NORMAL DISTRIBUTION FOR THE SAME TOTAL NUMBER OF CASES AND THE SAME STANDARD DEVIATION

the extent of the departure from normality of the scores in reading and in inverse ratio to their relationship with the scores in reading.

CRITERION FOR JUDGING COMPARISONS OF DISTRIBUTIONS

Here, then, we have a criterion for judging the comparisons which are presented in Diagrams II, III, and IV. It is the closeness of fit with the normal of an actual distribution of one set of measures when the number of cases is so small as to cause wide departure from the normal of the distribution of other sets of measures of the same cases. The matter may be stated in another way. When chance alone operates, comparatively few cases are needed to get an approach to the normal in distribution. When other influences operate with chance, many more cases must be taken in order to get an approach to normality, unless, as may be the exceptional happening, the distribution of the few in measures of the other influences also approaches normality. The distributions in Diagram IV serve as illustration.

The curves shown in Diagram IV represent the problem-difference scores of 195 cases. The cases are too few in number to be described, let us say, as a representative group in ability to read.

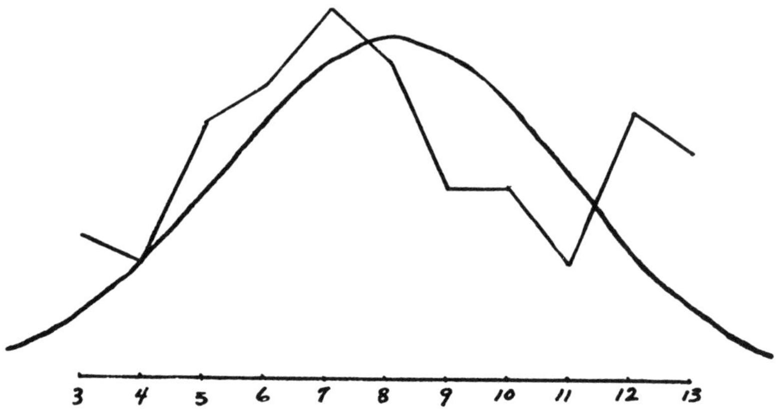

DIAGRAM V

Actual Distribution of the Burgess Reading Scores of the 195 Pupils Represented in Diagram IV Compared with the Corresponding Normal Distribution for the Same Total Number of Cases and the Same Standard Deviation

Diagram V shows how far the scores in reading of these 195 cases depart from the normal. If any relation existed between the problem-difference scores and ability to read, then the curve in Diagram IV would approach in form, either directly or in reverse, the curve shown in Diagram V. In spite of the fact that too few cases are represented to expect a normal distribution in the traits which are measured by the Collateral Tests, we find the problem-difference scores approximating the normal distribution. The operation of chance is demonstrated in Diagrams II, III, and IV, especially in Diagram IV. In Diagram IV, we find a closeness of fit of a few cases—too few to produce the normal distribution, if any discernible characteristics or abilities of the pupils operated in combination with chance. Our investigations have indicated that the abilities and qualities which are measured by the Collateral Tests and which are a part of or contributory to ability to attack ordinary problems do not affect the operation of chance in the determination of the point-and-problem-difference scores. Chance seems to play its part alone. The C–I point scores and the A–B scores are on the whole chance scores, and are not significant as individual scores. The only significance is in an accumulation of differences, such as have been explained by the slight superiority of the longer conventional problems, and of the shorter conventional problems when time is a factor.

The Relation Between Difference-Scores in Tests IV and V

Data are at hand to show in another way the influence of chance upon the difference-scores. Table 50 shows the correlation between the C–I problem scores in Test IV of 204 pupils

TABLE 50

The Relation Between the Problem-Difference Scores in Problem Test IV and the Problem-Difference Scores in Problem Test V of the Same Pupils

Difference Scores Test IV	Difference Scores—Test V								
	−2	−1	0	1	2	3	4	5	Total
4				1	1				2
3	3		5	1	6				15
2		5	6	7	5	2	3		28
1		4	9	14	8	1	1	1	38
0	3	8	18	17	9	8	3		66
−1	1	3	13	10	4	3		1	35
−2		1	6	5	1	2			15
−3		1	1	1	2				5
Total	7	22	58	56	36	16	7	2	204

$r = .02$

and the A–B problem scores in Test V of the same pupils. The table shows that whatever may be the difference score of a pupil in one test, his difference score may be the same or anything else in the other. The coefficient of correlation, $r = .02$, is hardly needed to call attention to the lack of relationship which is shown by the table.

The Relation Between Difference Scores in the Same Test

The influence of chance may be shown in another assembling of data. Table 51 shows the relation—rather, the lack of relation—between the C–I point scores in the AA, AS, AM, and AD pairs of problems of Problem Test I and C–I point scores on the remaining six pairs. The test was divided in this fashion, partly to facilitate the tabulation from the pupils' papers, but chiefly because such division separates the results into approximately equal halves. It may be seen by reference to Table 2 that in both the conventional and the imaginative problems the pupils did about as well, in point of number of problems solved, in the first four pairs as they did in the remaining six pairs.

If there is any particular virtue, even in individual cases, in either the one type of problem or the other, it might be expected to appear in the two halves of a single test, at least in a sufficient

TABLE 51

The Relation Between the C–I Point Scores in the AA, AS, AM, and AD Pairs of Problems and the C–I Point Scores in the Remaining Six Pairs of Problem Test I

(Papers with zero problem scores are not included)

Difference Scores, AA, AS, AM, AD Pairs	Difference Scores—Remaining Six Pairs											
	−5	−4	−3	−2	−1	0	1	2	3	4	5	Total
4					1			1				2
3		1		1	1	1	3		2			9
2				6	5	6	3	5	2	1	1	29
1			2	6	11	13	7	12	5	2	2	60
0		1	4	11	22	48	30	13	1		1	131
−1	1		5	8	11	20	18	4	5	3		75
−2	1	2	1	6	6	13	6	3	1	1		40
−3			1	2	4	1	5	3	1	1		18
−4				1					2		1	4
−5					1	1			1			3
Total..	2	4	13	41	62	103	72	41	20	8	5	371

$r = .04$

number of cases to make possible the recognition of its presence. Table 51 shows zero correlation between the point-difference scores in the two halves of the test. It confirms the fact that a difference score in one instance is no indication of any particular score in another. The only safe prediction is that if any difference is discoverable it will not vary greatly from zero. The direction of the variation is entirely a matter of chance.

Summary

1. The distributions of the point-difference scores in Problem Tests I and IV and of the problem-difference scores in Problem Test V fit closely the normal distributions.

2. Closeness of fit in difference scores is to be observed even when two few cases are represented to produce any appreciable approach to normality of such qualities as are measured by the Collateral Tests.

3. No correlation is to be found between the problem-difference scores of Problem Test IV and similar scores of Problem Test V.

4. The difference scores in the tests in question appear to be due to chance alone, and in individual occurrences are without significance.

CHAPTER IX

SUMMARY AND CONCLUSIONS

This chapter is intended to serve the double purpose of summarizing the findings which have been reported in the preceding pages, and of suggesting their probable explanations and values.

Reasons for the Investigations

If the reader of the opening paragraphs of this report were to take the trouble to examine the longer imaginative problems presented in the newer textbooks, his first question might probably be, Why not accept the longer problems for incorporation into the curriculum? The textbooks in which they are found are without doubt well organized, and cleverly put together. The longer problems are attractive in statement; at least, they strike the attention of the adult. They seem to accord with the prevailing tendency to make the work of the school as attractive as possible to the pupils. Moreover, they provide for the same variety in practice as do problems of the usual, conventional type. Indeed, in so far as their arithmetical features are concerned, they are not unlike the problems which the school has always used. Why not begin to use problems which are intended in their manner of statement to present an imaginative appeal?

To the foregoing question, the answer can be made that it is important to determine whether the imaginative type of problem is really the better type or not. If it should prove to be better than the conventional type of problem, then justification would be established for an immediate organization of effort to bring it into general use. On the other hand, if the longer, imaginative problems fail to establish their superiority, then their omission from textbooks could be argued on the double grounds of economy of lesson procedure, and of economy in cost of publication and in cost of books to pupils.

Method of the Investigations

The method of collecting the data for presentation and discussion in this report was the method of testing pupils upon pairs of problems. The responses of pupils to problems of the two types were collected and compared. The criterion for judging the superiority in merit of a given problem of either type over its paired associate of the other is the relative ease of solution.

Problem-Solving Activities Distinguished

In order to justify ease of solution as a criterion of merit, the purposes of the problem-solving activities were analyzed.

The learning of arithmetic in the early stages was defined, not as the mere memorization of number facts, but as the development of ideas of number and number relations. Such ideas result from the giving of attention in a systematic way to arrangement: first, to objects in actual arrangement; second, to objects arranged in thought; and, finally, to facts of arrangement when the objects are present only in imagination. "Problems" are thus defined as practice exercises, and "problem-solving" as the activity of recognition of ideas of arrangement made discoverable by the language which describes a familiar situation. The type of practice exercise which provides the readier recognition of arrangement to the point of diminishing returns is held to be of the greater merit as a means of training.

Such definitions of "problem" and "problem-solving" set off in sharp contrast the problem-solving activities of arithmetic in the earlier and later stages. In the early stages, pupils are called upon to give attention to problem practice exercises which deal with familiar situations during, and as a means of, learning the meaning and use of the fundamental operations. In the later stages, the attention of pupils is directed to problems in situations with which they are becoming familiar and in which, having already learned, they now use the fundamental operations.

Limitations of the Present Investigation

Because of the contrast, to which attention has just been called, the attempt has been made in the present investigation to avoid confusion in the issues by dealing only with the kind of "problem," or practice exercise, which is made use of in the early stages of arithmetic. Moreover, because such kind of problem has been considered throughout merely as a practice exercise, the tests which were used were made to involve only the four fundamental operations with whole numbers, and were administered to pupils of the fifth and sixth grades. By the time pupils have reached these grades, they have learned the meaning and use of such operations sufficiently well to recognize them in situations when they appear as single operations. That they have not learned their meaning and use perfectly and need further training and practice is evidenced by the fact that they are often in some doubt in recognizing these operations when they appear as double operations, as was the case in the problems chosen for the tests.

The Problem Tests

Three types of problem tests were used. (1) Problem Test I provided a comparison of practice exercises of the conventional type with practice exercises of the imaginative type. (2) Problem Test IV provided a similar comparison with the exception that the conventional problems were made longer in order to include in the wording of their statement suggestion as to the operations which were involved. For convenience the conventional problems in Test I were called the "shorter," and those in Test IV were called the "longer." In both tests the element of time was uncontrolled, all pupils being given ample time to try all the problems. (3) Problem Test V provided a comparison similar to the comparison provided in Test I, with the exception that it was administered with the element of time under control.

Problem Test II, which was a repetition of Myers' test, provided a type of comparison similar to that provided by Test I.

The Collateral Tests

For the purpose of providing sets of data for comparisons and analyses, four standardized tests were used in the investigations. From these, five sets of individual scores were obtained, namely, scores in ability to read short paragraphs, scores in intelligence, scores in carefulness of work, scores in arithmetical reasoning, and scores in the simpler operations in the fundamentals. The attempt was made to use these five sets of scores to explain the variations in the individual scores in the problem tests. Their use was not extended to all of the problem tests for the reason that they proved to be of little or no value in explaining variations in the results in the tests with which they were used.

Treatment of Results

1. The results which were obtained from the various problem tests were reported as difference scores, both in averages and in distributions of the groups tested. In individual cases, superiority of responses to problems of the one type over responses to problems of the other was indicated by the amount and sign of the difference.
2. From the problem scores and the point scores in two of the tests other scores were computed to show similarity in responses to the paired problems.
3. Both the difference scores and the similarity scores were grouped with the scores obtained from the Collateral Tests in order to discover possible evidences of relationship.
4. The influence of chance in producing difference scores in individual cases was studied through comparisons of their distributions with the normal, and by means of correlations of the difference scores made by the same pupils in two of the tests, and of the difference scores made by the same pupils in the two halves of the same test.

Conclusions

The results which have been reported justify the following conclusions with respect to the relative merits of problem practice exercises of the conventional and imaginative types in the earlier stages of the learning of arithmetic.

1. The negligible differences between the totals of responses to the two types of problems of Tests I and II indicate that pupils of the intermediate grades are neither hindered nor helped in their problem practice exercises by problems of the imaginative type, when no limits are imposed upon the amounts of time of the practice periods.

2. The expansion of conventional problems through extra wording designed to render them less difficult by explaining more fully and indicating more specifically than is usual the strictly mathematical elements of the situations may be expected to result, in certain instances, in a gain in correct responses.

3. A large percentage of pupils may be expected to be neither helped nor hindered by the longer conventional statements.

4. In the cases where help seems to be had, the amounts of helpfulness appear to be largely matters of chance.

5. The responses of pupils to the problems of the two types were more similar than different. Even many of the poorer pupils produced scores which were highly similar.

6. When the time of the practice periods is the same, more practice can be secured from problems of the conventional type than from problems of the imaginative type. The conventional problems appear to offer slightly superior opportunity for practice to all pupils.

7. Difference scores in individual cases can not be explained differences in ability to read, differences in intelligence, differences in carefulness of work, differences in arithmetical reasoning, or differences in ability in the fundamentals. Whatever the relation of these abilities and qualities to efficiency in practice with such problems as our tests presented, they appear to operate with equal effectiveness, whether the problems are short or long, conventional or imaginative.

8. Differences in the individual scores in similarity of response to problems of the two types may be explained only in part by differences in the abilities and qualities just referred to. Pupils who ranked high in these abilities and qualities as a rule made high scores in similarity, whereas the pupils who ranked low made both high and low scores in similarity.

9. Low intelligence, carelessness of work, and poor arithmetical reasoning appeared to result in chance, sporadic, haphazard successes more frequently than low ability in reading and in the fundamentals.

10. The difference scores which resulted from the various problem tests appeared to be due to chance alone. Differences in individual cases are without significance.

Implications of the Findings

While it is not within the province of this report to speculate upon the methods of instruction in arithmetic, a suggestion or two in this respect which seem to be implied in the findings may be excused.

One may observe among teachers a tendency to contrast what they term the concrete with what they term the abstract, greatly to the disadvantage of the latter. In other words, that which is concrete shall be magnified, and that which is abstract shall be avoided. To this end, there is included with the exercises dealing with "problems" in arithmetic much reference by the teacher to the concrete phases of the total situation which surround or are a part of the abstract number relations involved. Often, part of the exercises is discussion of these so-called concrete elements. Moreover, as pointed out in the first chapter of this report, a tendency to introduce into the printed problem practice exercises of pupils the type of imaginative elements which relate to the concrete may be noted.

No doubt, the pupil who is confronted with an arithmetic problem to solve must get into the field of his attention enough of the concrete elements of the total situation to give shape and meaning to the number relations involved. To be sure, there must be present in consciousness enough of the so-called concrete to provide the materials, that is, the situation, from which the number relations are to be abstracted. But, as the second chapter of this report has indicated, it is the fact of number relation, of arrangement, which should command the attention. Upon this element of the total situation the pupil must concentrate. It may be that too much emphasis upon the other elements will serve as a means of distraction. Our findings indicate that when the emphasis is attempted by the printed exercise, the pupil is not helped.

It appears, moreover, that whatever of the total situation may be necessary to provide the background which determines or explains the number relations, the pupil must reconstruct for himself. In other words, the pupil must be trained to do his own thinking, or, in the language of the problem of this investigation, to provide for himself whatever of imaginative elements is needed.

What the printed statement provides, while it may not hinder, does not help.

Is the problem void of imaginative elements necessarily dull? The act of dealing with the situation requires the neglect of all the extra imaginative elements which may be thrown in. Since it is the fact of arrangement which must command the attention, any other appeal to the pupil's interest is wasted effort. Our findings indicate that when a pupil has formed an acquaintance with a given type, or combination, of number relations, for example, the situation of a double relation of addition and division, he is able to recognize it, and thus becomes better acquainted with it, in whatever manner it may be presented.

In order to learn it is necessary that the pupil give attention to essentials and neglect non-essentials. It would seem to be the part of wisdom for teachers and textbooks, whenever they attempt guidance and assistance, to emphasize essentials and to minimize non-essentials.

Need for Further Study

There is implied in the present report, more particularly in the theoretical considerations of the second chapter than elsewhere, the need for further investigations and study. It may be that the school is inclined to emphasize in too great detail the situations out of which the problems of the early stages of arithmetic are taken, and to emphasize too little the situations which provide the problems of the later stages of arithmetic. It may be that the school, in the manner of its organization of arithmetic and in the methods of its presentation of the subject, has failed to distinguish with sufficient clearness the study of number relations and the study of situations. These statements must, however, for the present, be considered as conjectural. Such distinctions as they indicate may be more imaginary than real. But so long as the distinctions remain with us as open questions with their consideration in the realm of conjecture, the school will be unable to proceed with complete confidence in the type of activities it recommends to its pupils.

The present study throws some light upon the handling of familiar situations in the early stages of arithmetic. It does not touch the question of how situations shall be dealt with in the later stages of arithmetic.

In the study of a situation in the later stages of arithmetic, for example, square measure, shall pupils be confronted with problems in square measure as soon as the topic is reached and as introductory to the study of the topic? Or shall square measure be presented, considered, discussed, studied with reasonable thoroughness before any problems are presented for attack and solution? When the problems are reached, shall they be stated tersely, or shall the particular situation of each problem be presented in considerable detail? Shall the problems all be alike in their manner of statement, or shall those at the beginning be presented with more elaborate descriptions than those toward the end?

The present investigation has sufficient justification in that it serves to introduce and encourage attack upon the question as to the place and purpose of the problem practice exercise in the different stages of the process of learning arithmetic.